South-Western

WESTWIND PROPERTIES:

PROBLEM SOLVING USING COMPUTER APPLICATIONS

William G. Perry, Jr., Ph.D.
Professor of Accounting and Computer Information Systems
Western Carolina University
Cullowhee, North Carolina

Reviewer:

Rene Zuniga
Mission High School
Mission, Texas

JOIN US ON THE INTERNET
WWW: http://www.thomson.com
EMAIL: findit@kiosk.thomson.com   A service of I(T)P®

South-Western Educational Publishing
an International Thomson Publishing company I(T)P®

Cincinnati • Albany, NY • Belmont, CA • Bonn • Boston • Detroit • Johannesburg • London • Madrid
Melbourne • Mexico City • New York • Paris • Singapore • Tokyo • Toronto • Washington

Copyright © 1998
by SOUTH-WESTERN EDUCATIONAL PUBLISHING
Cincinnati, Ohio
ALL RIGHTS RESERVED

The text of this publication, or any part thereof, may not be reproduced or transmitted in any form or by any means, electronic or mechanical, including photocopying, recording, storage in an information retrieval system, or otherwise, without the prior written permission of the publisher.

ISBN 0-538-71727-0
Printed in the United States of America

I(T)P®
International Thomson Publishing Company

South-Western Educational Publishing is a division of International Thomson Publishing, Inc. The ITP trademark is used under license.

Managing Editor	Janie F. Schwark
Project Manager	Dave Lafferty
Marketing Manager	Kent Christensen
Design Coordinator	Ann Small
Development and Production	Thompson Steele Production Services

Dedication

To Karen—the love of my life

CONTENTS

Part I — Word Processing Skills

Beginning Word Processing Skills — 1
- Activity 1 Creating a Document — 2
- Activity 2 Retrieving and Modifying a Document — 3
- Activity 3 Modifying Words and Phrases — 4
- Activity 4 Using Help and File Operations — 4
    - *Other Activities and Questions* — 5
    - *Key Words and Phrases* — 6

Intermediate Word Processing Skills — 7
- Activity 1 Saving Files in an Alternate Format — 7
- Activity 2 Specialized Retrieval of Documents — 8
- Activity 3 Creating a Boilerplate Document — 9
- Activity 4 Creating a Newsletter — 11
- Activity 5 Block or Highlighting Selected Text — 13
    - *Other Activities and Questions* — 13
    - *Key Words and Phrases* — 14

Advanced Word Processing Skills — 15
- Activity 1 Creating Documents with Graphics — 15
- Activity 2 Preparing a Form Letter — 16
- Activity 3 Creating a Document with a Table — 17
- Activity 4 Merging a Data File — 19
- Activity 5 Inserting Clip Art — 20
    - *Other Activities and Questions* — 21
    - *Key Words and Phrases* — 22

Part II — Spreadsheet Skills

Beginning Spreadsheet Skills — 23
- Activity 1 Creating a Basic Spreadsheet — 24
- Activity 2 Retrieving and Modifying a Spreadsheet — 26
    - *Other Activities and Questions* — 27
    - *Key Words and Phrases* — 28

Intermediate Spreadsheet Skills — 29
- Activity 1 Designing a Spreadsheet Template — 30
- Activity 2 Modifying Templates and Selected Output — 31
- Activity 3 Creating a Simple Spreadsheet Graphic — 32
- Activity 4 Inserting and Deleting Rows and Columns — 32
    - *Other Activities and Questions* — 33
    - *Key Words and Phrases* — 34

Advanced Spreadsheet Skills — 35
- Activity 1 Modifying Cell Labels — 35
- Activity 2 Linking Spreadsheets — 37
- Activity 3 Creating a Customized Graph — 38
- Activity 4 Modifying Spreadsheet Output — 39
    - *Other Activities and Questions* — 39
    - *Key Words and Phrases* — 40

contents continued on page vi

Beginning Database Skills — 41
- Activity 1 Designing and Using a Database — 41
- Activity 2 Adding and Deleting Files — 43
- Activity 3 Changing the Database's Structure — 44
  - *Other Activities and Questions* — 45
  - *Key Words and Phrases* — 46

Intermediate Database Skills — 47
- Activity 1 Using Key Fields and Calculating Fields — 47
- Activity 2 Viewing Data and Correcting Errors — 49
- Activity 3 Renaming a Field — 51
- Activity 4 Editing and Backing Up a Database — 53
- Activity 5 Querying a Database and Using Logical Operators — 55
  - *Other Activities and Questions* — 55
  - *Key Words and Phrases* — 56

Advanced Database Skills — 57
- Activity 1 Customizing a Data Entry Form — 57
- Activity 2 Customizing a Report — 59
- Activity 3 Developing an Interactive Application — 60
- Activity 4 Importing, Querying and Exporting — 61
  - *Other Activities and Questions* — 63
  - *Key Words and Phrases* — 64

Part III Database Skills

Beginning Presentation Graphics Skills — 65
- Activity 1 Creating a Presentation Graphics Slide — 66
- Activity 2 Using Clip Art — 67
- Activity 3 Modifying a Slide and Printing a Copy — 68
  - *Other Activities and Questions* — 69
  - *Key Words and Phrases* — 70

Intermediate Presentation Graphics Skills — 71
- Activity 1 Creating a Master Slide — 71
- Activity 2 Creating a Properly Aligned Series of Slides — 72
- Activity 3 Modifying Text Objects and Creating a Slide Show — 73
- Activity 4 Editing a Slide Show — 74
  - *Other Activities and Questions* — 74
  - *Key Words and Phrases* — 75

Advanced Presentation Graphics Skills — 76
- Activity 1 Creating Slide Show Special Effects — 76
- Activity 2 Altering Slide Images and Importing Graphics — 77
- Activity 3 Creating a Runtime Version Slide Show — 78
  - *Other Activities and Questions* — 79
  - *Key Words and Phrases* — 80

Part IV Presentation Graphics Skills

contents continued on page vii

Part V Operating System Skills

Beginning Operating System Skills — 81
- Activity 1 Understanding Disk, Directory and Screen Operations — 82
- Activity 2 Using Essential File Operations — 83
- Activity 3 Accessing On-line Help — 83
  - *Other Activities and Questions* — 84
  - *Key Words and Phrases* — 85

Intermediate Operating System Skills — 86
- Activity 1 Copying a File — 86
- Activity 2 Creating Specialized Storage Space — 87
- Activity 3 Working with a Group of Files — 88
- Activity 4 Deleting Files and Specialized Storage — 88
- Activity 5 Adjusting the Time and Date — 89
  - *Other Activities and Questions* — 89
  - *Key Words and Phrases* — 90

Advanced Operating System Skills — 91
- Activity 1 Backing Up and Restoring Files — 92
- Activity 2 Modifying a File's Contents — 92
- Activity 3 Sorting Files — 93
- Activity 4 Altering the System's Configuration — 93
- Activity 5 Changing File Attributes — 94
  - *Other Activities and Questions* — 95
  - *Key Words and Phrases* — 96

Part VI Graphical User Interface Skills

Beginning Graphical User Interface Skills — 97
- Activity 1 Activating and Using the GUI — 97
- Activity 2 Opening and Closing GUI Applications — 98
- Activity 3 Becoming Familiar with the GUI — 98
- Activity 4 Using the GUI's On-line Help — 99
  - *Other Activities and Questions* — 99
  - *Key Words and Phrases* — 100

Intermediate Graphical User Interface Skills — 101
- Activity 1 Manipulating Windows — 101
- Activity 2 Formatting a Disk — 102
- Activity 3 Using the Clipboard — 103
- Activity 4 Using a GUI Accessory — 103
  - *Other Activities and Questions* — 104
  - *Key Words and Phrases* — 105

Advanced Graphical User Interface Skills — 106
- Activity 1 GUI Customization and Multiple Windows — 106
- Activity 2 Controlling Output with the GUI — 107
- Activity 3 Altering the GUI System Font — 108
  - *Other Activities and Questions* — 108
  - *Key Words and Phrases* — 109

contents continued on page viii

Appendix A	**Word Processing Skills**	111
	Record Sheets	113
Appendix B	**Spreadsheet Skills**	115
	Record Sheets	117
Appendix C	**Database Skills**	119
	Record Sheets	121
Appendix D	**Presentation Graphics Skills**	123
	Record Sheets	125
Appendix E	**Operating System Skills**	127
	Record Sheets	129
Appendix F	**Graphical User Interface Skills**	131
	Record Sheets	133

PREFACE

People learn best when they can apply new skills and knowledge in a realistic context. Westwind Properties: Problem Solving Using Computer Applications *follows this learning format.*

The purpose of this book is to supplement the instructional materials in your computer science or business education courses. The exercises in the book are geared toward giving the student a realistic format in which to apply what he or she has learned about his or her particular operating systems, word processor, spreadsheet, database and presentation graphics software. Each unit challenges you to apply what you have learned to solve realistic problems.

The author's goal is to help you gain confidence in your ability to utilize modern information-processing tools to solve problems similar to those you will encounter on-the-job after completing your studies.

This text is divided into six parts. The first four parts cover word processing, spreadsheet, database and presentation graphics software. Sections five and six cover operating systems and graphical user interfaces.

Each part contains three levels of competencies or skills for students to master: Beginning, Intermediate and Advanced. Students assume the role of a real estate company employee and they apply their skills and knowledge to solve realistic problems. These problems move from simple to more complex.

Premises of the Exercises

Westwind Properties, a real estate and property management company, has just hired you. The owner/manager, Ms. Blaire Jones, asked you to join her staff because of your computer knowledge. Indeed, she expects you to handle all of the "computer-related matters" at Westwind Properties.

Ms. Jones has some computer knowledge, but she is eager to learn more and to have you help the entire staff.

Approach each learning activity with the idea that you are attempting to please your employer.

Part I
Word Processing Skills

Beginning Word Processing Skills

Word processing software has virtually replaced traditional document preparation in business and government. A person who is capable of creating, editing, and distributing his or her own documents is recognized as being more productive.

Many managers require employees to submit a first draft of a document for review before distributing a final copy. This document will most likely be returned with changes. Indeed, you might decide to revise the document's contents or appearance. For example, when reading a document for the second time you might discover that you used the same word over and over again. To avoid this repetition, you could use the word processor's electronic thesaurus to examine a list of alternative words with similar meanings. Also, you could use the search and replace feature to change each occurrence of a word or phrase to a different one throughout an entire document.

There may be times when you need to share a document with another individual who has a different word processor. Many modern word processors allow you to save a document so that it can be read by different word processors. In this hands-on lab unit you will be able to explore and apply all of these skills from within the context of an employment setting.

Preview

Modern word processing software makes it easy to create, save and change electronic documents. The appearance, size and location of text can be routinely altered.

Using word processors to prepare a document is made more efficient by the use of a thesaurus and spell checker. The finished product can be saved using the word processor's own file format or by selecting a different choice.

Activity 1 Creating a Document

In this activity you will:

- Create and save a document that is **fully justified**
- Center a title on a document
- **Highlight** text
- Change default text **fonts**
- Change a font **style**
- Use a **spell checker**
- Produce a hard copy printout of the document

Situation:

Ms. Blaire Jones, the owner and manager of Westwind Properties, has asked you to write a memo introducing yourself to the staff. However, she wants to review the first draft of the memo before you distribute it.

Step 1—Enter the text shown below for your memo. Use your word processor's **default font style** with a 12-point size. Highlight the word "Memorandum", give it bold face type and center it. Fully justify the remainder of the text. Use a **default tab stop** to align the inside address.

Memorandum

(Current Date)

To: The Staff
From: (Insert Your Name)
Subj: Greetings!!

My name is (insert your name). Ms. Jones has hired me to help you more productively use desktop computers. Even though I am new to the game of real estate, I understand computers and I'm looking forward to meeting you. I am thrilled to be a part of the team.

Hopefully, I'll be able to provide you with some ideas as time goes by. Please don't hesitate to let me know if you have any questions.

Step 2—After you have entered the text, run your word processor's spell-checking tool and correct any errors that are found.

Step 3—Save the memo to your disk and name the file **GREET**. Accept the **default file extension** name that is used by your word processor.

Step 4—Print out a copy of your memo and exit the word processor.

Required Output: A hard copy printout of the memo, free of spelling errors and a disk containing the file **GREET**.

Activity 2 Retrieving and Modifying a Document

In this activity you will:

- Retrieve a document from disk
- Change font sizes
- Use the word processor's **insert** and **type over mode**

Situation:

Ms. Jones has read your memorandum and has returned it with some minor changes, one of which is to increase the type size.

Step 1—Insert your data disk into the appropriate drive and open the file titled **GREET**. Change the point size for the entire document from 12 to 14.

Step 2—Shown below is the memo with Ms. Jones' changes. Rather than re-entering the entire memo, make the changes in the wording that are indicated with italics using the insert and type over mode. Don't forget to make the word "thrilled" appear in a **bold style.**

Step 3—Run your word processor's spell checker once again and make any corrections that might be identified.

Memorandum

(Current Date)

To: The Staff
From: (Insert Your Name)
Subj: Greetings!!

Dear Staff: Please allow me to introduce myself. My name is (insert your name). Ms. Jones has hired me to help *us all become more familiar with using* desktop computers. Even though I am new to the game of real estate, *I feel confident I'll be able to help you with using the microcomputer.*

I am **thrilled** to be a part of the team *and shall look forward to meeting each of you over the next couple of weeks.* However, please don't hesitate to drop by my office in the meantime if you have any questions *or suggestions.*

Step 4—Save the revised document to your data disk. If asked if you wish to overwrite or replace the existing file, respond "yes".

Step 5—Print out a copy of the revised memo.

Required Output: A hard copy printout of the revised memo.

Activity 3 Modifying Words and Phrases

In this activity you will:

- Modify a document using the word processor's **thesaurus** feature
- Modify a document using the **search** and **replace** feature of the word processor

Situation:

After proof reading the revised memo, you've decided that you don't like some of your word choices and want to make additional changes.

Step 1—Open the file named GREET. Highlight the word "thrilled" and activate your word processor's thesaurus feature. Select a new word from the list of alternatives that are proposed and replace "thrilled". Make certain that the new word you select is still bold faced.

Step 2—You've decided to replace each occurrence of the term "desktop computers" with "microcomputers." Place the **cursor** at the beginning of the document. Activate your word processor's **search and replace** feature. In the appropriate manner, indicate that you wish to replace each occurrence of the phrase "desktop computers" with the word "microcomputers."

Step 3—Save the revised version of GREET to your data disk and print out a hard copy of your revised document.

Required Output: A hard copy printout of the revised document.

Activity 4 Using Help and File Operations

In this activity you will:

- Use the Save As option to store an existing document in **ASCII text format** under a different file name
- Use the word processor's **on-line Help** feature
- Delete a document from storage

Situation:

You anticipate that the need will arise to share files among various individuals. Saving a document in an ASCII file format is one way that different word processors can share files. You decide to experiment with saving a document as an ASCII file.

Step 1—Open the memo named GREET from your data disk and access your word processor's **on-line Help** feature. Read the instructions on how to save a document in a different **file format.**

Step 2—Using the Save As feature of your word processor, save the document GREET as an ASCII text file named MEMO_1. Delete the version of your document named GREET.

Step 3—Print out a hard copy of the text file titled MEMO_1.

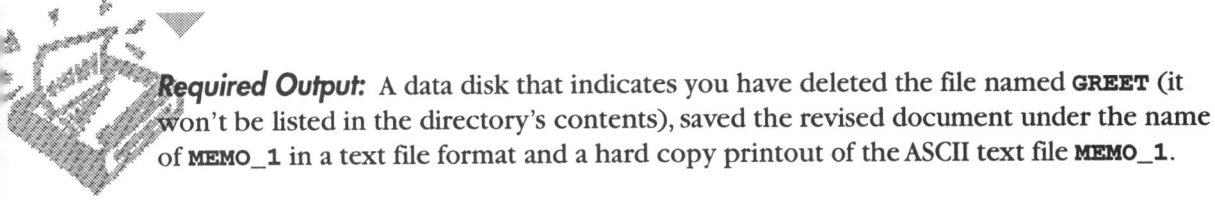

Required Output: A data disk that indicates you have deleted the file named **GREET** (it won't be listed in the directory's contents), saved the revised document under the name of **MEMO_1** in a text file format and a hard copy printout of the ASCII text file **MEMO_1**.

Other Activities and Questions

1. Do you think that a document saved in ASCII text file format can be used by any word processor? Why? Why not?

2. How does the ASCII file printout **MEMO_1** differ from the word processor's printout of the file named **GREET**?

3. Can words be added to the spell checker feature of your software? If so, how?

4. Can words be added to the thesaurus feature of your software? If so, how?

5. Can a word processor's spell checker normally determine the correct spelling of a person's last name? Why or why not?

6. List the advantages of modern word processing over that of more traditional methods.

7. Can additional fonts be added to your word processor?

8. Write a letter to a friend, relative or loved one. Produce a printout and mail it to them.

Key Words and Phrases

ASCII text file format—American Standard Code for Information Interchange is a standard method of representing letters, numbers and special characters.

bold style—A feature contained in most word processors that allows for text to be displayed in a shade that is darker than regular text style.

cursor—A visible indicator displayed on the screens of most software, that indicates the relative position where input will be received.

default file extension—The letter or letters automatically attached to the main file name by the word processor.

default font style—The initial style in which the word processor displays text.

default tab stops—The horizontal positions, automatically assigned by the software, where the cursor stops.

file format—Any one of a number of "file" types in which software can store documents.

fonts—The graphic form in which letters, numbers and characters are displayed.

fully justified—Text in a document which is displayed with even left and right margins.

highlight—A method of selecting text so that specific operations may be performed upon it.

insert mode—The text entry method that allows for letters, numbers or characters to be inserted at any point in an existing document without destroying existing data.

on-line Help—The feature contained on most modern software that provides help files that can be displayed on the computer's screen.

search and replace feature—A tool that searches an existing document for one particular phrase and replaces any occurrences with another word or phrase specified by the user.

spell checker—A word processing feature that compares the accuracy of each word in a document with the words contained in the software's dictionary.

style—The appearance of a text's font, either regular, bold or italics.

type over mode—A method of changing the letters, numbers or characters in a word processing document that "writes over" or destroys what is displayed.

thesaurus—A word processing tool that allows the user to select a word and review a list of alternative words that have a similar meaning.

Intermediate Word Processing Skills

Co-workers often need to share documents, which can be a problem if people are using different word processing software and hardware systems. In this lab unit, you will have an opportunity to save a word processing document in a different word processor's file format so that a co-worker can use it.

Companies often need a form or a blank document that they can reuse. In this unit, you'll create a boilerplate invoice that can be used over and over.

The appearance of word processing documents is just as important as the content. Some documents are used to attract attention and communicate important ideas. In this unit, you'll use a word processor to create a marketing document.

Preview

Word processors are powerful tools and may be used by virtually everyone in the organization. Many word processors have a feature that allows a document to be stored in a different **word processor's format.**

Another powerful feature of a word processor provides for the creation of a document that can be used repeatedly. A user can create a document with headings and blank spaces to be filled in later. The blank document can be saved and then recalled. Once the data is entered, the document can be saved under a different file name.

Word processors also provide for a variety of formatting that make it possible to create documents with a sophisticated look.

Activity 1 Saving Files in an Alternate Format

In this activity you will:

- Save a document in a different word processing file format

Situation:

Ms. Jones tells you that she frequently uses her computer at home to prepare and print out documents for the office. She uses a different word processor than the one being used in the office but she wants to be able to exchange files.

activity 1 is continued on page 8

She asks you if this is possible. You explain that it is and decide to create a test document on the office's word processor. You decide to save the file on disk using the file format of Ms. Jones' word processor so that she can see if she can use it on her computer at home.

Step 1—Open your word processor and access the on-line Help feature and review how to save a word processing document in a different file format.

Step 2—Type in the memo shown below.

(Current Date)

Memorandum

To: Mr. Ralph Peterson
From: Ms. Blaire Jones, Broker
Subj: The Deep River Property

The owner of the Deep River property, Harold Lieberman, telephoned me this morning. I discussed with him the nature of the offer you were preparing to make.

Mr. Lieberman indicated that he would be open to considering the flexible financial terms that would be a feature of your offer.

Indeed, he indicated to me that he would appreciate receiving such offer from you by the end of the week.

Please let me know if I may be of assistance.

Step 3—Save the memo to a disk in your word processor's default file format. Name the file **LET1**. Next, save the document in the format of Ms. Jones' word processor. (*Your instructor will specify the format.*) Name this file **LET2**.

Required Output: A printout of the disk's contents which should show the two documents, **LET1** and **LET2**.

Activity 2 Specialized Retrieval of Documents

In this activity you will:

- Retrieve a document from a drive and directory other than the default drive and directory
- Edit a document that was saved in a different word processing file format

Situation:

Ms. Jones comes in and tells you that she was able to retrieve the document named **LET2** on her home computer. She gives you back the disk and asks you to see if you can open **LET2** after it was saved on her computer and still make changes to it on the office computer.

Step 1—Insert the disk in your computer's disk drive and open the document named **LET2**.

HELPFUL HINT

Read the section of your word processor's on-line Help feature that relates to retrieving a document which has been saved in a different file format.

Step 2—After you have **LET2** on the screen, insert the following title, *Ms. Jones' WP Document*, on the page.

Step 3—Save the document in your word processor's file format and print out a copy of the document.

Required Output: A printout of Ms. Jones' **LET2** document that contains the new title.

Activity 3   Creating a Boilerplate Document

In this activity you will:

- Use **tab stops**
- Change default margins
- Create a boilerplate document

Situation:

Ms. Jones asks you to create a document that people in the office can use as a generic invoice for clients who have reserved one of the seasonal rentals. She wants the invoice to look the same regardless of who prepares the bill. You realize what Ms. Jones needs is a **boilerplate** document.

Step 1—Read the following special instructions:

SPECIAL INSTRUCTIONS: Reduce the **default margins** by five spaces on the right and five spaces on the left.

Step 2—Enter the document that is shown on the following page. Evenly space the headings "Item", "Description" and "Rate." Use the text formatting options of your word processor to create different size fonts and styles on the invoice.

activity 3 is continued on page 10

HELPFUL HINT

Make sure to use tab stops below the column headings so that the data inserted in each column will be properly aligned under the heading.

```
                              INVOICE
    Date:                                        Invoice #: _____
    To:
    From:  Ms. Blaire Jones, Manager
           Westwind Property Management and Realty
           3101 Westwind Drive
           Boulder, Colorado  80306
    ─────────────────────────────────────────────────────────────────
           Item              Description                    Rate
    ─────────────────────────────────────────────────────────────────

                             Subtotal                    _____
                             Sales Tax                   _____
                             TOTAL                       _____
```

Step 3—Save your boilerplate document as **INV_PERM**.

HELPFUL HINT

INV_PERM should **always** be saved blank. Use **INV_PERM** as an empty form. When you want to save an invoice that contains data, use the Save As feature of your word processor.

Step 4—Print out a copy of **INV_PERM**.

Step 5—Next, use the blank invoice **INV_PERM** to enter the set of data shown below:
     *Note that the client has rented the unit for 4 weeks.

```
Invoice #:   8783

Samantha Martinez
104 Sunset Blvd.
Ft. Lauderdale, FL  33449

Items:    2 Bdrm Condo
          Weekly Linen Service

Description:   Astrid Flower – 4 weeks

Rate:     $1,700.00 for the condo per week
          $35.00 per week for the Linen Service

Sales Tax:   Apply the standard 9% rate
```

Page 10

Step 6—Use the "Save As" feature of your word processor to save the invoice under the name **IN8783**.

Required Output: A printout of a blank invoice or template and a printout of invoice **IN8783** which contains the actual data.

Activity 4 Creating a Newsletter

In this activity you will:

- Center column headings
- Use your word processor's basic text formatting tools
- Use your word processor's **page numbering system**
- Use the **line spacing** feature of your word processor
- Change the orientation to your word processor's output

Situation:

Ms. Jones decides to publish a newsletter for Westwind Properties and asks you to produce it using the word processor. She gives you a list of design items to follow when creating the first draft.

Step 1—Read the following special instructions:

SPECIAL INSTRUCTIONS:

1. There should be three columns of text in the newsletter.
2. The tile of the newsletter should be centered and in bold face type.
3. The column headings should be centered.
4. The third paragraph of the newsletter's introduction should be in boldface type.
5. The text in the columns should be fully justified.
6. Each page of the newsletter should be numbered.
7. The beginning of each paragraph should be indented and the spacing between lines should be more than "single" and less than "double".
8. Ms. Jones wishes to examine one copy in a **portrait orientation** and one in a **landscape orientation.**

Step 2—Enter the text below and on the following pages for the newsletter.

The Westwind

Introduction

　　Westwind Properties is proud to send you its first issue of <u>The Westwind</u>. As a former customer, we hope you will find the information contained in this newsletter to be of use to you.
　　We have made a special effort to brief you on the array of services that we continue to offer property owners, prospective clients, renters and buyers. We've also added a few new services that we think you'll find interesting.

activity 4 is continued on page 12

We hope you enjoy The Westwind. Write us and let us know what you think about our newsletter. Your ideas and suggestions for improvement are welcomed.

Property Management News

Westwind Property Management has done something that is almost unheard of in this world. We've reduced prices! Our usual property management fee has been 15% but effective immediately we have reduced the percentage figure to 12%. That's a 3% savings for our clients.

We've been able to automate some of our property management services and have begun using a competitive bidding process with independent housekeeping contractors. Prior to this point in time we maintained our own housekeeping staff. We believe that we'll be able to provide better service to our owners in this manner. We have passed along the savings to you.

Yes, Colorado has it all. In the summer, tourists flock from the hot cities to our beautiful mountains. They are able to relax and enjoy cool mountain breezes. For the more adventuresome, hiking, rafting, fishing and camping are all there for the taking.

All of these tourists need a place to stay and, like you, they'd enjoy renting and staying in a single family residence rather than a commercial motel.

We manage the renting and maintenance and you enjoy the tax and investment benefits. The unit is yours when you want it and can earn you money when you don't.

One of our specialists would be happy to discuss our owner and rental program with you. Stop by our office on your next visit or give us a call.

Rentals Are on the Rise!!

Westwind Properties specializes in renting units, owned by individuals, to high-quality clients who are visiting our beautiful area.

We took a look at the numbers last week and were surprised to learn that gross rental receipts exceeded 2.5 million dollars last year. That represents a significant increase over last year's 1.9 million.

All factors being equal, the rental market is growing. The number of units becoming available and the number of inquiries and walk-ins are on the rise.

Competitive Bidding Yields Savings

We have maintained our own housekeeping staff in the past. This required us to devote significant staff resources to manage the housekeeping employees. We also incurred expenses relating to benefits and worker compensation.

We made the decision to request competitive bids from professional housekeeping companies in the area and, based upon the results of the bid, have decided to independently contract with professional housekeeping firms to clean and prep rentals when renters vacate and new clients enter. So far we are pleased with the results.

As mentioned earlier in this newsletter, we have cut our property management fees by 3% (from 15% to 12%) and have decided to pass on the savings to our owners.

Please let us know if you receive any negative feedback on the changeover to independent housekeeping services. We'll closely monitor the quality of their work from this end.

We've Enhanced Our Computer Services

Westwind Properties is proud to announce that we have hired (Place Your Name Here) to run our computer services. (Your First Name) is a whiz on computers and has already helped us become more efficient. Indeed, (Your First Name) prepared this newsletter and is responsible for making sure that you each receive a copy.

Stop by when you are in town and say hello to (Your First Name).

Step 3—Spell check the newsletter and make any necessary corrections.

Step 4—Save the document as **NEWS**.

Required Output: A printout of the newsletter in a portrait orientation and a printout of the newsletter in a landscape orientation.

Activity 5 Blocking or Highlighting Selected Text

In this activity you will:

- Use the block command
- Delete text that has been blocked

Situation:

Ms. Jones reviewed the two versions of the newsletter and decided that she prefers the portrait layout. She asks you to remove the last paragraph of the *Introduction* section and print out the final copy.

Step 1—Open the document named **NEWS**.

Step 2—Highlight or **block** the last paragraph, delete it and save the revised file.

Step 3—Print out a copy of the newsletter.

Required Output: A printout of the newsletter in a portrait format without the paragraph inviting comments from readers.

Other Activities and Questions

1. What other types of documents do you think could be stored in boilerplate form and then recalled for data entry?
2. Create a single-page newsletter that describes a recent activity in which you participated.
3. In how many different word processing file formats can a document be stored on your word processor?
4. Is there more than one way to change margins on your word processor?
5. Why would someone want to use tab stops?
6. Should everyone else in Ms. Jones' office now be able to use a template file that you create? If so, why? If not, why?
7. List the text formatting tools that are available in your word processor.

Key Words and Phrases

block—The highlighting of a specific area of text in a document.

boilerplate—A stored document in which certain data is permanent or fixed (i.e. column headings, title, etc.) and variable data can be entered.

default margins—The margin settings automatically supplied with a word processor's new page.

landscape orientation—Output that is horizontally oriented.

line spacing—The amount of vertical space between lines of text in a word processing document.

page numbering system—That feature of a word processor that assigns page numbers to a multi-page document.

portrait orientation—Output that is vertically oriented.

tab stops—A word processing feature that allows for the pre-programming of "stops" along the horizontal line.

word processor's file format—A unique file type in which a document is stored that varies from word processor to word processor.

Advanced Word Processing Skills

There is a broad range of capabilities included in modern word processing software. Many include a number of special effects features, such as a draw program, a clip art library and the capability to create tables and special effects that can enhance documents.

Also, most word processors possess the capability to produce mass mailings.

In this lab unit, you'll have an opportunity to use a number of your word processor's advanced features.

Preview

Word processing software can be used to meet a business' routine document preparation needs as well as to perform some desktop publishing functions.

Most of today's word processors also have the capacity to **merge** a **form letter** with a **data file.** A data file with hundreds or even thousands of names and addresses can be merged with a form letter. The result is an "original" letter to each person in the data file.

The fact that word processing software is so powerful can raise employer expectations for both the office and professional staff. Knowing how to utilize a word processor's more advanced features is a distinct advantage for any employee.

▶ Activity 1   Creating Documents with Graphics

In this activity you will:

- Create a word processing document that contains a graphic
- Use your word processor's drawing tools to create a simple image
- Create a document that contains a **header** and **footer**
- Add **borders** and **shading** to a portion of a document

Situation:

Even though you lack the education of a professional artist, Ms. Jones asks you to create an advertising flier that includes a clock face to accompany the theme of the flier that "Time is Running Out!" She wants the clock face to contain numbers and be positioned in the upper left-hand corner of the document.

▼

activity 1 is continued on page 16

Step 1—Open your word processor and start a new document by entering the header "Time Is Running Out!" This header should be surrounded by a box and shaded to appear as though it is three dimensional. If necessary review your on-line Help on how to create and center a header and use borders and shading.

Step 2—Use the following text for the flier.

> That's right, "Time Is Running Out" if you plan on making a reservation for this year's skiing season. The supply of quality rentals is limited and every year we have to turn away customers because the vacancies aren't available when people want to visit.
>
> There's only one way to beat the rush and assure yourself and your family of a place to stay while you are enjoying the slopes this winter. Make your reservations now!
>
> Give us a call. Our eager staff is waiting to serve you.

Step 3—Review how to use your word processor's drawing program. Experiment with drawing circles, lines and placing numbers within a circle.

Step 4—Create the clock face and insert it in the upper, left-hand corner of the document.

Step 5—Save the document as CLOCK.

Required Output: A printout of the advertising flier you created. The header should appear in a shaded box. The clock face should appear in the upper left-hand corner of the document.

Activity 2   Preparing a Form Letter

In this activity you will:

- Use the word processor to prepare a form letter

Situation:

Ms. Jones needs to send the same letter to a number of clients. In the past, she's prepared a letter to a single customer and printed it out. Then, she deleted the customer's name and address and re-entered another customer's name and address and printed out the letter again. She repeated the same process until all of the names on the customer list had been used. This is very tedious and she asks if there is a more efficient way to send an original letter to customers.

You explain that you can create a form letter and merge it with a list of clients, eliminating the need to manually re-enter names and addresses for each printout of the letter.

Step 1—Access your word processor's on-line Help feature. Read how to prepare a form letter to be merged with a data file.

Step 2—Create the form letter. Enter the following text. Make sure you include the footer at the bottom of the letter. Save the file as `formlet`.

Text for the Main Document:

> Current Date:
>
> Name of Customer
> Address
> City, State Zip
>
> Dear Customer:
>
> You have been one of our most loyal clients over the years. That's why we want you to have the first chance to rent one of our brand-new flagship properties.
>
> All of them have a hot tub and sauna. They also have a southern orientation and skylights. All of them are within five minutes walking distance of the lifts and they are all luxuriously decorated.
>
> Let us know which one to reserve for you.
>
> Sincerely,
>
> Blaire Jones
> Manager
>
> (Insert a footer stating: "Your One Stop Rental Services Company")

Step 3—Print out a copy of the letter for Ms. Jones to review.

Required Output: A printout of the form letter.

Activity 3   Creating a Document with a Table

In this activity you will:

- Create a document that contains a **table**

activity 3 is continued on page 18

Situation:

Ms. Jones reads the form letter you prepared for her. She decides that the letter should be more descriptive. Therefore, she asks you to prepare a table that lists the new rental units and to insert it into the form letter.

Step 1—Review your on-line Help on how to create a table to be included in a letter. Retrieve **formlet**, if necessary.

HELPFUL HINT

Add the text that is shown in italics.

Step 2—Modify the first paragraph of the form letter to read as shown below.

> Current Date:
>
> Name of Customer
> Address
> City, State Zip
>
> Dear Customer:
>
> You have been one of our most loyal clients over the years. That's why we want you to have the first chance to rent one of our brand-new flagship properties. *Shown below is a listing of them:*

Step 3—Create a table to be placed in your form letter using the following data.

Table to be Inserted in the Document:

> *Listed below are the units and their rates:*
>
> | Unit # | Size | Daily | Weekly | Monthly |
> |--------|--------|----------|------------|------------|
> | 6087 | 3 brdm | $275.00 | $1,800.00 | $6,950.00 |
> | 6088 | 4 bdrm | $320.00 | $2,000.00 | $7,600.00 |
> | 6089 | 2 bdrm | $265.00 | $1,700.00 | $6,500.00 |
> | 6090 | 3 bdrm | $285.00 | $1,850.00 | $7,200.00 |
> | 6091 | 3 bdrm | $250.00 | $1,500.00 | $5,850.00 |
> | 6092 | 4 bdrm | $320.00 | $2,100.00 | $7,800.00 |

Step 4—Insert the new paragraph in your form letter as shown on the following page.

Each of the above rentals fall under the category of "can you believe it?"

All of them have a hot tub and sauna. They also have a southern orientation and skylights. All of them are within five minutes walking distance of the lifts and they are all luxuriously decorated.

Let us know which one to reserve for you.

Sincerely,

Blaire Jones
Manager

(Insert a footer stating: "Your One Stop Rental Services Company")

Required Output: A hard copy printout of the form letter which includes the table.

Activity 4 Merging a Data File

In this activity you will:

- Create a data file to be merged with a form letter

Situation:

Ms. Jones approves of your revised form letter that includes the table. Now you need to create the data file that is to be merged.

Step 1—Read the on-line Help materials on how to create a data file to be merged with a form letter.

Step 2—Enter the data below and on the following page into the data file.

List of Customers to Receive Letter:

Ralph W. Woodson
P.O. Box 87
Lake Placid, FL 33872

Linda Carson
872 4th Street North
Athens, Georgia 40482

activity 4 is continued on page 20

Donald Berington
219 Circle Drive
Bend, Oregon 97330

Mack Langston
489 32nd Street
East Grand Forks, MN 58202

Step 3—Save your file, if prompted to do so, under the name `custinfo`.

Step 4—Review how to produce hard copy printouts of your merged form letter.

Step 5—Print out your form letters.

Required Output: A hard copy printout of four letters each addressed to a different customer.

Activity 5 Inserting Clip Art

In this activity you will:

- Select a graphic from your word processor's clip art library and insert it into a document

Situation:

Ms. Jones asks you to prepare a simple document that contains an image from your word processor's clip art library. She wants to see how professional the word processor's clip art looks.

Step 1—Read your on-line Help about how to use clip art.

Step 2—Create a new document with the title "Example of a Clip Art Image." Center the title.

Step 3—Preview the clip art library. Select an image and insert it into the document. Below the image, describe the picture.

Step 4—Save the file under the name `clipart` and print out a copy for Ms. Jones.

Required Output: A printout of the image you chose along with the required title in the header and a brief, written description of the image.

Other Activities and Questions

1. Can your word processor access any clip art library?
2. Can your word processor perform math within the table that has been created in the document?
3. Can you import an image into your word processor's draw program? If so, what would be the advantage of doing so?
4. Can you place regular text in header and footer space?
5. Can you place a graphics image in the header and footer space?
6. Is the data file associated with a form letter a database? If so, why? If not, why?
7. Create a data file that includes the names and addresses of your family and friends. Create a form letter to send to them. Produce a letter to each person on your list.

Key Words and Phrases

borders—A box that may be placed around text or a graphic.

data file—A file that contains records.

footer—Space that appears at the bottom of each page of a document.

form letter—A letter which may be merged with a data file and replicated.

header—Space that appears at the top of each page of a document.

merge—The joining of appropriate information in a data file with a document.

shading—The background coloring associated with a box.

table—A special word processing format that includes rows and columns.

Part II
Spreadsheet Skills

Beginning Spreadsheet Skills

Spreadsheets are used to organize numbers, conduct "what if" financial analysis, and to prepare reports such as budgets or profit and loss statements. Countless hours of tedious calculation and re-calculations are saved by the use of spreadsheets and spreadsheet templates. Analysts can adapt virtually any spreadsheet to their own proven financial analysis techniques.

You can create your own spreadsheet or you can use an existing template and enter variable data. After you have created and tested a spreadsheet template you can use it over and over again. Many businesses create a spreadsheet that employees can use each month. Employees retrieve the template, which contains labels and formulas, and insert variable data. The "Save As" option is then used to store the current spreadsheet under a different name.

An employee who knows how to construct a spreadsheet can create and easily use this valuable financial analysis tool. In this lab unit you will create, save, modify and print a spreadsheet.

Preview

Spreadsheets organize text and numbers in a two-dimensional table made up of columns and rows. The intersection of a **row** and column is called a **cell.** Cells may contain three different types of data: a numerical **value,** a **label** (text) and **formulas.**

Numerical values may be formatted as whole numbers, real numbers, currency or in other formats. Labels are descriptive titles and formulas specify how the numerical value for a cell is to be calculated.

Determining the proper layout and format of a spreadsheet is very important.

Activity 1 Creating a Basic Spreadsheet

In this activity you will:

- Enter **labels** and **values** in a **spreadsheet cell**
- Construct and enter **formulas** in spreadsheet cells
- Format numerical data as currency with two decimal places
- Save a spreadsheet to disk
- **Block** an appropriate **range** in a spreadsheet for printing
- Print out a spreadsheet
- Change default column widths

Situation:

Ms. Jones asks you to create a **cash flow analysis** spreadsheet that can be used by all of the sales associates to help them analyze the income producing potential of rental properties. Ms. Jones supplies you with the outline of the spreadsheet **template** that is shown below and on the following page.

Step 1—Open your spreadsheet application. Select the option from the menu that allows you to create a new spreadsheet.

Step 2—Study the layout of the Cash Flow Analysis spreadsheet template before entering any data. Insert the title or header *Cash Flow Analysis*.

Step 3—Next, enter the labels on the left-hand margin of the spreadsheet. Increase the width of the **column** containing the labels so that it is wide enough to fit the longest label. Format this column as text and italic.

Step 4—**Format the cells** in the second and third columns to contain currency values with two decimal places.

Step 5—Define the formulas for the **cells** that are to hold calculated values.

Step 6—Save the spreadsheet **template** as `TEMP_SS`.

Step 7—Produce a hard copy printout of the spreadsheet template.

HELPFUL HINT

The cells in the Cash Flow Analysis sheet that need to contain formulas are displayed in bold face. Also, make sure that the widths of the columns to hold currency values are wide enough.

Cash Flow Analysis Sheet

Income
Scheduled Rental Income $xxx,xxx.xx
    Plus Other Income $xx,xxx.xx
Total Income **$xxx,xxx.xx**

    Less Bad Debts $xx,xxx.xx

Gross Operating Income: $xxx,xxx.xx

List of Operating Expenses:
Accounting and Legal	$xx,xxx.xx
Advertising, Licenses and Permits	$xx,xxx.xx
Property Insurance	$xx,xxx.xx
Property Management (7%) of Gross Operating Income	$xx,xxx.xx
Payroll	$xx,xxx.xx
Taxes	$xx,xxx.xx
Repairs and Maintenance	$xx,xxx.xx
Services	$xx,xxx.xx
Electricity	$xx,xxx.xx
Gas & Oil	$xx,xxx.xx
Water	$xx,xxx.xx
Garbage	$xx,xxx.xx
Miscellaneous (3%) of Gross Operating Income	$xx,xxx.xx

Less Total Operating Expenses	$xxx,xxx.xx
Net Operating Income	$xxx,xxx.xx
Less: Total Annual Debt Service	$xx,xxx.xx
Cash Flow Before Taxes	$xxx,xxx.xx

Step 8—Enter the values shown below into your spreadsheet template so you can test the accuracy of your template and its formulas.

Scheduled Rental Income	$39,840.00
Other Income	$757.84
Total Income	(To be Calculated)
Bad Debts	$1,035.00
Gross Operating Income	*(To Be Calculated)*
Accounting and Legal	$750.00
Advertising, Licenses and Permits	$1,500.00
Property Insurance	$3,650.00
Property Management (7% of Gross Operating Income)	(To Be Calculated)
Payroll	$2,400.00
Taxes	$3,740.00
Repairs and Maintenance	$323.20
Services	$400.00
Electricity	$3,563.23
Gas & Oil	$2,897.98
Water	$1,459.00

activity 1 is continued on page 26

Garbage	$650.00
Miscellaneous (3% of Gross Operating Income)	(To Be Calculated)
Total Operating Expenses	(To Be Calculated)
Net Operating Income	(To Be Calculated)
Total Annual Debt Service	$17,560.00
Cash Flow Before Taxes	(To Be Calculated)

Step 9—Double check the accuracy of your spreadsheet once you have entered the data. Save the spreadsheet to your data disk under the name of **ANALYSIS**.

HELPFUL HINT

Use the Save As option, otherwise, you will overwrite the template file named **TEMP_SS**.

Step 10—Highlight or block the appropriate range on your spreadsheet and print out a copy.

Required Output: A hard copy printout of your spreadsheet template named **TEMP_SS**, and a hard copy printout of the spreadsheet file named **ANALYSIS**.

Activity 2 Retrieving and Modifying a Spreadsheet

In this activity you will:

- **Copy the contents of a spreadsheet cell** to another cell
- Insert a row and a column into an existing spreadsheet
- Retrieve a spreadsheet
- Access and use the spreadsheet's on-line Help feature

Situation:

Ms. Jones reviews the hard copy printout of the spreadsheet named **ANALYSIS**. She notices that a new expense category, named "Supplies," needs to be added. You recognize that you must insert a row in both the spreadsheet named **ANALYSIS** and in the template, **TEMP_SS**.

Also, to make the spreadsheet more attractive you are asked to insert a column between the labels and the first column of numbers and to add a new label several lines below the last label called "PROFIT OR LOSS".

HELPFUL HINT

Use your spreadsheet's on-line Help feature to review how to insert a row and column.

Step 1—Open the **ANALYSIS** spreadsheet. Position the cursor on the cell containing the expense category "Electricity". Insert a row.

Step 2—After you have successfully inserted a row, add an expense category named "Supplies" in Column A. Enter the value of $853.18 in the cell to the right of the label.

Step 3—Edit the formula for "Total Operating Expenses" to include the value for "Supplies".

Step 4—Now, let's make your spreadsheet more attractive. Insert a column between the expenses column and the first column of figures.

Step 5—Add a label several spaces below the last row of your existing spreadsheet in the left-hand margin, named "PROFIT OR LOSS." Access your spreadsheet's on-line Help feature and review the steps required to copy the contents of one cell to another. Copy the contents of the cell containing the Cash Flow Before Taxes value and place it into the cell immediately to the right of the cell labeled "PROFIT OR LOSS."

Step 6—Save your modified **ANALYSIS** spreadsheet. If your program asks if you wish to overwrite the original spreadsheet, respond "yes." Produce a printout of the new **ANALYSIS** spreadsheet.

Step 7—Lastly, make the changes in your blank spreadsheet template named **TEMP_SS** and produce a hard copy printout.

HELPFUL HINT

Be sure that you have edited the appropriate formula on both files, **TEMP_SS** and **ANALYSIS**.

Required Output: A hard copy printout of the revised **ANALYSIS** spreadsheet, and a hard copy printout of the revised blank template named **TEMP_SS**.

Other Activities and Questions

1. What is the purpose of creating a spreadsheet template?
2. Identify and describe at least three other ways that spreadsheets could be used in business.
3. Describe how arithmetic operators and parentheses are used to build spreadsheet formulas.
4. Create a spreadsheet template into which you can enter you own personal weekly or monthly budget.

Key Words and Phrases

block—A group of highlighted cells.

cash flow analysis—A financial report that indicates the cash left over after expenses.

cell—The intersection of a row and a column in a spreadsheet.

column—Vertical delineation of space on a spreadsheet usually labeled with a letter.

copy the contents of a spreadsheet cell—Duplicating contents of a spreadsheet cell into another.

format the cells—Specifying how data is to appear in cell.

formulas—A mathematical expression that is one type of value that can be entered into a cell.

labels—Alphabetic titles entered into a spreadsheet's cell.

range—A group of cells indexed by the upper left and the lower right-most cell address.

row—Horizontal space on a spreadsheet.

spreadsheet cell—A specific address on a spreadsheet that consists of both a column and row reference.

template—A spreadsheet that contains pre-saved labels or constants that is routinely retrieved for the entry of variable data.

values—Generally, numeric data or a formula.

Intermediate Spreadsheet Skills

Spreadsheets are typically used to analyze revenue versus expenses. Frequently, there is a need to redo such a spreadsheet at regular intervals (weekly, monthly or quarterly). Rather than recreating the spreadsheet each time, a template that contains fixed labels and formulas may be used. When the spreadsheet template is opened, variable data is entered. The resulting spreadsheet can be saved under a different name, thus preserving the original template for future use.

Spreadsheets also have **pre-defined functions** that can save time. One spreadsheet function, SUM, makes it possible to total the values contained in a range of cells rather than each cell having to be listed individually.

Spreadsheet data may also be attractively displayed. The values in a spreadsheet's cell may need to be centered within a column or placed on either the left or right hand margin. These text formatting tools are also provided in spreadsheets. In addition, spreadsheet data can be presented in a chart or graph. Familiarity with the spreadsheet software's data formatting and graphic capabilities makes it possible to present spreadsheet data in an effective manner.

You will have an opportunity to create a spreadsheet template, use functions, format data, and create a spreadsheet graphic in this lab unit.

Preview

Financial recordkeeping is essential for the successful operation of a business. The tasks include totaling revenue from various sources as well as performing other calculations.

Spreadsheets can be created that are unique to a business. Standard labels and formulas can be used to create a template. When a spreadsheet template is recalled and new data is entered, the spreadsheet may be saved. An accurate spreadsheet template can save a tremendous amount of time.

Spreadsheet templates, like all spreadsheets, begin with a blank page. Reoccurring labels are entered in appropriate cells and formulas are defined. The spreadsheet template may be tested with a sample set of data but saved blank.

When a spreadsheet template is recalled for use, variable data can be entered. The "Save As" option may then be used to save the new spreadsheet under a different name and the template is preserved.

Graphs or charts may also be created using certain values in a spreadsheet.

Activity 1   Designing a Spreadsheet Template

In this activity you will:

- Create a spreadsheet template
- Format and **align** spreadsheet data
- Use pre-defined spreadsheet functions

Situation:

Ms. Jones wants to create a spreadsheet template that tracks the income generated by the rental units on a **quarterly** basis.

Step 1—Open your spreadsheet software. Insert the spreadsheet title and column heads that are shown below. Use a different font for the title and column heads. **Format the data** to be contained in the "Weekly Rental Rate", "Gross Revenue", "Comm. Rate" and "Comm. Earned" columns as **currency data.** Center the column headings. Use your spreadsheet's default column width and center the labels within the column. Enter the unit numbers and commission rates. Format the commission rates as percentages.

Rental Income and Commissions Analysis Sheet

Unit Number	Weekly Rate	Weeks Rented	Gross Revenue	Comm. Rate	Comm. Earned
xxxx	$xxx.xx	xx	$x,xxx.xx	xx%	$xxx.xx
	GRAND TOTALS:		$xx,xxx.xx		$x,xxx.xx

Step 2—Enter the formulas needed in the spreadsheet to compute the **gross revenue** and **commissions earned** for each unit. Format all the dollar values as currency.

HELPFUL HINT

Be sure to provide a sufficient number of rows for data in your spreadsheet template.

Step 3—Review the on-line Help feature on how to use the pre-defined SUM function. Apply the SUM function to obtain the GRAND TOTALS for both Gross Revenue and Comm. Earned.

Step 4—Save the spreadsheet template as `QT_TEMP`. Print out a copy of the template.

Step 5—Enter the data shown on the following page for the First Quarter's spreadsheet. These values provide the data needed to calculate commissions.

HELPFUL HINT

You may need to widen certain spreadsheet columns in order for them to be able to hold the calculated values.

Quarter One

Unit 3234, with a 10% commission rate, rented for 9 weeks at $350.00 per week
Unit 3239, with a 15% commission rate, rented for 6 weeks at $485.00 per week
Unit 3328, with a 12% commission rate, rented for 4 weeks at $735.00 per week
Unit 3335, with a 15% commission rate, rented for 2 weeks at $415.00 per week
Unit 3339, with a 20% commission rate, rented for 5 weeks at $345.00 per week

HELPFUL HINT

Check the accuracy of your spreadsheet formulas with a calculator.

Step 6—Save Quarter One's spreadsheet under the name **QT1_INC**. Print out a copy of **QT1_INC**.

Required Output: A printout of the properly formatted spreadsheet template named **QT_TEMP**, and a printout of the spreadsheet **QT1_INC** containing the data for Quarter One.

Activity 2 Modifying Templates and Selected Output

In this activity you will:

- Modify an existing spreadsheet template
- Print out a selected portion of a spreadsheet

Situation:

Just as you finish the First Quarter spreadsheet, Ms. Jones tells you that Rental Unit 3328 has been sold and that a new commission rate of 15% has been established. Now you have to modify the spreadsheet template. You will need to print out a copy of the revised template **QT_TEMP** and retrieve the spreadsheet file **QT1_INC** and make the necessary changes related to Unit 3328.

Ms. Jones also asks you to print out a listing of just the rental unit numbers so that she has a handy list of the units available for rent.

Step 1—Open the spreadsheet template named **QT_TEMP**. Change the formula for Rental Unit 3328 to reflect the new commission rate. Test the new formula using the first quarter's revenue numbers. After you're sure the formula is correct, delete the data and save the spreadsheet template. Print out a copy of the revised template.

Step 2—Retrieve Quarter One's file. Make the appropriate changes. Produce a printout of **QT1_INC** and save the file.

activity 2 is continued on page 32

Step 3—Review how to print out only a portion of a spreadsheet in your spreadsheet's on-line Help. Print out the rental unit numbers only.

Required Output: A printout of the revised spreadsheet template `QT_TEMP`, a printout of the revised spreadsheet `QT1_INC` showing the modified commission rate for Unit 3328, and a printout of the column that contains only the numbers of the rental units.

Activity 3   Creating a Simple Spreadsheet Graphic

In this activity you will:

- Create a simple bar graph from spreadsheet data

Situation:

You decide to impress Ms. Jones with your spreadsheet's graphics capabilities. Utilize your spreadsheet's **quick graph** or **chart** feature to create a bar graph that shows the commissions for each rental unit.

Step 1—Read how to use the quick graph or chart feature using your spreadsheet's on-line Help feature.

Step 2—Open the First Quarter spreadsheet. Create a bar graph that shows the commissions for each rental unit in comparison to the total. Title the bar graph *Commissions on Unit Rentals Compared to Total Commissions* and print out a copy.

The rental unit numbers should appear on the left side of the graph and the commission amounts should appear on the bottom line of the graph.

HELPFUL HINT

Each rental unit's commission will be represented with a bar. Total Commissions will have a bar of its own. You may need to hide certain columns and rows to get the desired results.

Step 3—Save the file `QT1S_INC`.

Required Output: A printout of the bar graph.

Activity 4   Inserting and Deleting Rows and Columns

In this activity you will:

- Insert and delete a row of spreadsheet data
- Insert a column in an existing spreadsheet

Situation:

Ms. Jones has found two mistakes in the spreadsheet template **QT_TEMP**. She inadvertently left out Rental Unit 3240. Unit 3240 should show a commission rate of 15% and that it rented for 3 weeks in Quarter One for $375.00. Second, Unit 3339 has been sold and is no longer available as a rental from your company. She also wants to add a "Percent of Total" column to the spreadsheet template.

Step 1—Review how to delete and insert a row and how to insert a column using your spreadsheet's on-line Help feature.

Step 2—Delete the row of data for Unit 3339 which has been sold.

Step 3—Insert the basic data for the new unit, 3240, between the rows for Units 3239 and 3328.

Step 4—Insert a new column between the "Gross Rental" and "Comm. Rate" columns and title it "Percent of Total". Each cell in this new column should contain the result of dividing Gross Revenue by the GRAND TOTAL of all of the gross revenue figures.

HELPFUL HINT:

Formulas must be contained for each cell in the "Percent of Total" column.

Step 5—Save the revised spreadsheet template as **QT_TEMP**. Print out a copy of the revised spreadsheet template.

Step 6—Open Quarter One's spreadsheet **QT1_INC**. Modify the spreadsheet according to the changes made in the template. Save the modified spreadsheet, **QT_INC**, to disk and print out a copy.

Required Output: A printout of the modified spreadsheet template **QT_TEMP**, and a printout of Quarter One's modified spreadsheet **QT1_INC**.

Other Activities and Questions

1. Design a spreadsheet template that lists personal checks to be written. Allow for general categories under which each expenditure could be categorized (i.e. entertainment, clothing, gas, etc.). Use a function to generate the total. Save your template.

2. Develop a list of fictitious checks. Retrieve your spreadsheet template and enter the data. Save your spreadsheet containing the data under a different name. Design a quick graph that displays how much you spent in each category when compared to the total spent.

3. How many predefined functions does your spreadsheet have?

4. Select a predefined function, other than one you've already used. Create a spreadsheet that uses the function you selected.

Key Words and Phrases

align—Selective placement of data contained within a spreadsheet's cells.

commissions earned—A percentage of sales that is paid as a fee.

currency data—Numerical data which represents money.

format the data—Altering the appearance of values in the spreadsheet's cells.

Gross Revenue—The amount of revenue obtained, in this exercise, when rate is multiplied times the number of weeks.

pre-defined functions—Functions such as Average and Sum that are contained as part of the spreadsheet software.

quarterly—A period of time, three months, by which business activities are frequently summarized and measured.

quick graph—The default graphing feature contained in some spreadsheet software packages.

Advanced Spreadsheet Skills

Spreadsheet software has advanced features that allow users to customize their work. Column parameters such as width, numeric formatting, different typefaces for labels and headings to enhance the appearance of spreadsheets can be changed. Sections from existing spreadsheet documents can be cut and pasted into a different spreadsheet. Many spreadsheets even support linking with another spreadsheet.

You can create graphs that pictorially express the spreadsheet's results or you can hide columns so that you can print out a subset of the spreadsheet. You can also export data from the spreadsheet to other documents, such as a word processing document.

You will have an opportunity in this lab to work with a number of a spreadsheet software's advanced features.

Preview

Spreadsheet users can change default column widths to accommodate larger values. In addition, the manner in which cell values are formatted can be revised. For example, a numeric value could be formatted as currency, a whole number or a real number with several decimal places. Also, some spreadsheets have features that allow you to change the color of a cell or group of cells as well as to shade and outline cells and include borders.

The typeface, size and style of fonts that can be used for cell labels can also be changed in the cells of spreadsheets. Generally, the text in spreadsheet cells can be altered using the same fonts as word processors. This is a particularly powerful feature when designing reports.

Many spreadsheets are capable of creating graphics which are **dynamically linked.** When data is changed on one spreadsheet, the graph is automatically changed, too. There are a number of specialized tools that are included with spreadsheets.

Activity 1   Modifying Cell Labels

In this activity you will:

- Change cell label fonts
- Change cell label parameters

activity 1 is continued on page 36

Situation:

You were surprised to learn that Westwind Properties actually owns eleven rental units. All of the revenue obtained from renting these units is retained by the company. But these rental units are not incorporated into the existing information system. You recommend to Ms. Jones that you create a separate spreadsheet listing the operating revenue, expenses and net operating income for each of the eleven units for the current year. Ms. Jones agrees.

Step 1—Open your spreadsheet application and review the on-line Help on how to change fonts, column widths and numeric formats.

Step 2—Create a spreadsheet that includes a column for the name of the rental unit, its number, its weekly rental rate, the number of weeks it was rented, the operating expenses, and a column for gross and net income. Make sure you format the weekly rate, operating expenses, gross revenue and net income columns as currency. Enter the data shown below.

HELPFUL HINT

The "Gross Revenue" and "Net Income" columns will not contain values until you define the appropriate formulas.

Raw Spreadsheet Data:

Unit Name	Unit Number	Weekly Rate	Weeks Rented	Operating Expenses
Blue Ridge Run	1101	$375.00	35	$4,987.50
Ribbon Cove	1102	$420.00	32	$5,107.20
Branson Branch	1103	$385.00	30	$4,389.00
Lilac Row	1104	$415.00	32	$5,046.40
Laurel Hill	1105	$435.00	33	$5,454.90
Bubbling Creek	1106	$450.00	38	$6,498.00
Oak Bend	1107	$435.00	37	$6,116.10
Pine Valley	1108	$385.00	34	$4,974.20
Green Lake	1109	$390.00	41	$6,076.20
Lofty Mist	1110	$430.00	38	$6,209.20
Stone Hill	1111	$410.00	33	$5,141.40

Step 3—Add a title, *Westwind Properties Rental Unit Net Operating Income Totals*. Use a different typeface other than the default. Be sure to center the title horizontally.

Step 4—Create the formulas for the columns that are to hold gross revenue and net income figures for each unit. Also, add a row for Totals and define a formula for the gross revenue, operating expenses and net income columns.

HELPFUL HINT

The "Totals" row should calculate the total gross revenue, operating expenses and net income.

Step 5—Adjust column widths, where necessary, to accommodate the largest field of data plus five additional spaces. Center the data in the weeks rented column.

HELPFUL HINT

You may discover a column width is too narrow for a calculated value only after you have entered the data. You will receive an error message. Simply expand the column.

Step 6—Save the spreadsheet as **SSREV** on your disk.

Required Output: A printout of the completed spreadsheet.

Activity 2 Linking Spreadsheets

In this activity you will:

- **Link spreadsheets**

Situation:

Ms. Jones asks you to create a second spreadsheet that just lists the yearly totals for gross revenue, operating expenses, and net operating income. She plans on using the second spreadsheet as a page in the annual report.

You explain to Ms. Jones how you can link one spreadsheet to another and recommend doing so for this second spreadsheet. She tells you to proceed that way.

Step 1—Read how to link spreadsheets and how to edit formulas in your on-line Help.

Step 2—Create a new spreadsheet, like the one shown below, with rows that contain labels for "Gross Revenue", "Operating Expenses" and "Net Operating Income". Add a fourth row titled "Percent of Total" and define a formula for the appropriate cell which yields the quotient of dividing operating expenses total by gross revenue total. Save the new spreadsheet as **ANLRPT**.

Westwind Properties Annual Operating Income Figures	
Gross Revenue	(Linked Cell)
Operating Expenses	(Linked Cell)
Net Operating Income	(Linked Cell)
Percent of Total	(Calculated Cell)

activity 2 is continued on page 38

Step 3—Link the gross revenue, operating expenses and net income from the **SSREV** spreadsheet to the appropriate cells in the new spreadsheet.

HELPFUL HINT

Your spreadsheet software may utilize different terminology to describe "linking". Ask for help from your instructor, if necessary.

Step 4—Print out a copy of the spreadsheet **ANLRPT** for Ms. Jones.

Ms. Jones reviews the printout of the new spreadsheet and realizes that she gave you the wrong formula for Percent of Total. The correct formula should divide net operating income by gross revenue.

Step 5—Edit or correct the formula associated with the Percent of Total cell. Save the revised spreadsheet and print out a copy.

Required Output: A printout of the original linked spreadsheet (the one containing the error), and a printout of the revised spreadsheet.

Activity 3   Creating a Customized Graph

In this activity you will:

- Create a **customized spreadsheet graphic**
- Cut and paste data

Situation:

Ms. Jones asks you to prepare a customized pie chart of the spreadsheet containing the annual operating income figures for gross revenue, operating expenses and net income.

Step 1—Review your on-line Help on how to create a customized pie chart for a spreadsheet.

Step 2—Retrieve the spreadsheet file named **ANLRPT**. Create a pie chart for this spreadsheet. The pie should contain three slices—gross revenue total, operating expense total, and net income total. Preview your pie chart.

Step 3—Add labels for each of the pie's three slices.

Step 4—Open the spreadsheet named **SSREV**. Cut the title from the **SSREV** spreadsheet. Close the spreadsheet.

HELPFUL HINT

When asked if you want to save the changes to your **SSREV** spreadsheet, respond negatively.

Step 5—Paste the title on the spreadsheet named **ANLRPT** that contains the pie chart. Save the spreadsheet **ANLRPT**. When you are finished print out a copy of the pie chart.

Required Output: A printout of the pie chart with external labels.

Activity 4 Modifying Spreadsheet Output

In this activity you will:

- Sort data in a spreadsheet column
- **Export or print spreadsheet data to a file**
- **Hide a column**
- Use the Save As feature of a spreadsheet

Situation:

Ms. Jones asks you to produce an alphabetized list of the properties owned by Westwind Properties and a disk file that shows the names of the units and the revenue each generated. You decide to use the first spreadsheet, **SSREV**, to construct the alphabetized list.

Step 1—Open the first spreadsheet that you saved named **SSREV**.

Review in your on-line Help how to sort data, hide columns and export data.

Step 2—Sort the Unit Name column in ascending order. Hide all the other columns and change the title on the spreadsheet to Units Owned. Then use the Save As feature of your program to save the spreadsheet under the name **SSREV2**. Print out a copy of the spreadsheet that shows the list of properties owned by Westwind Properties.

Step 3—Now open the spreadsheet file named **SSREV**. Prepare a new spreadsheet that only shows the column with the rental unit names and the gross revenue generated. Save the spreadsheet under the name of **REVENUE**. Save the file for export.

HELPFUL HINT

Most spreadsheets can save output to a disk file. Read your on-line Help or ask you instructor for assistance, if necessary.

Required Output: A printout of the spreadsheet **SSREV2** that lists the names of the rental units, and a printout of the disk's contents that show an exported file named **REVENUE**.

Other Activities and Questions

1. Import the data file you created in Activity 4 into a blank spreadsheet.

2. Why would a person want to import data into a spreadsheet?

3. How many different graphics types are available in your spreadsheet?

4. What are the major spreadsheet software packages that are available on the market?

5. How many different ways may numeric data be formatted in your spreadsheet format? Why would one or the other be used?

Key Words and Phrases

customized spreadsheet graphic—A graph that follows an original rather than a default format.

dynamically linked—A feature which causes any change in a spreadsheet's values to be transferred to any graph associated with the original spreadsheet.

exporting spreadsheet data to a file—Directing the software to send a file "outside" the original program.

hide a column—The spreadsheet feature that allows you to make selected columns invisible.

linking spreadsheets—A feature of spreadsheet software that allows a portion(s) of one spreadsheet to be mathematically joined to another.

Part III
Database Skills

Beginning Database Skills

Computers are best at storing data and retrieving records in a useful form. Businesses use database software to store data related to customers, inventory, sales transactions, employees and more. Database application software lets a user define what data is to be stored, how it should be formatted, and how it should be displayed when retrieved.

Learning how to store strategic information for processing and later retrieval is the essence of all data processing. In this lab you will design and save a database.

Preview

A **database** is a collection of records (for example, the names of customers). A **record** in a database consists of fields (for example, a customer number or a last name).

A database must first be defined or created. Each field in a database's record must be named and have its **data type** specified. When the database is defined, data may be input using the software's **default data entry format.** Individual records may then be retrieved.

Specialized reports, based upon a database's contents, can be generated. However, a simple printout of the database's contents can be obtained by using the software's **default report format.**

▶ ### Activity 1   Designing and Using a Database

In this activity you will:

- Design a database

activity 1 is continued on page 42

- Define a field
- Specify a field's data type
- Save a newly created database
- Enter data using a **default format** to a data disk
- **Sort records** in a database
- Generate a report using a default format

Situation:

Ms. Jones asks you to create a database of her sales associates that would include their names, addresses, home telephone numbers, real estate license numbers and ages.

HELPFUL HINT

Be certain that you specify, when defining the database, that you want it saved to the data disk.

Step 1—Start your database application. Decide upon the **field** names and their data or field type. Make sure that the associates' first, middle and last names are each treated as a separate field. Consider the remaining fields in the database to be **character** or alphabetic data.

HELPFUL HINT

Although there are special data types associated with various database software applications, data that isn't used for calculations may be considered to be alphabetic or character data type.

Step 2—When you have defined the fields in the database, enter the data below and on the following page using the software's default data entry format.

Raw Data:

Marilyn Watson
101 West 42nd St.
Ft. Lauderdale, FL 33940

Age: 59

Home Phone: 305-887-3366

License #: 343-844-AL-3171

Larry Candler
P.O. Box 97
Dania, FL 34399

Age: 42

Home Phone: 305-774-3343

License #: 388-877-FD-3755

Mary Beth Wilson
8732 East 15th Avenue
Perine, FL 33978

Age: 39

Home Phone: 305-871-3789

License #: 778-343-GB-8788

James Lanier
9th St. NW
Del Ray Beach, FL 33488

Age: 29

Home Phone: 305-338-8831

License #: 878-879-XT-7788

Catherine Ruth Jones
788 Wentworth Drive
Ft. Lauderdale, FL 33877

Age: 38

Home Phone: 305-701-7801

License #: 887-237-FS-7899

Ricky Schultz
596 Booth Drive
Boca Raton, FL 34789

Age: 29

Home Phone: 305-778-8347

License #: 792-347-JR-7800

Step 3—Next, sort your database by last name in **ascending order.**

Step 4—Print out a hard copy of the sales associate information using the default report format supplied by the software.

Required Output: An alphabetized report for the sales associates sorted in ascending order.

Activity 2 Adding and Deleting Files

In this activity you will:

- **Insert a new record or row** in an existing database
- **Delete a record** or row in an existing database

Situation:

Ms. Jones warned you that turnover is high among the sales associates. And sure enough, on the morning after you completed entering the data into your database, one sales associate (James Lanier) resigns and another is hired.

Step 1—Delete the record for James Lanier from the database.

Step 2—Add a record for the newly hired sales associate using the data shown below.

Ms. Marian Schmel
8931 SW 32nd St.
Boynton Beach, FL 33479

Age: 48

Home Phone: 813-344-3847

License #: 778-338-FG-7024

Step 3—Resort the database in ascending order.

activity 2 is continued on page 44

Step 4—Generate a hard copy printout of the revised database using the default report format.

Required Output: A printout of the database in the software's default format.

Activity 3 Changing the Database's Structure

In this activity you will:

- **Modify the structure** and contents of an existing database
- Insert a column or field in an existing database
- Remove a column or field from an existing database
- Access and use the database software's on-line Help feature

Situation:

You become aware that "Age", as a field in an employee database, could possibly be used against the company in an age discrimination suit. You mention your concern to Ms. Jones and she agrees that "Age" should be deleted. Also, during your conversation with Ms. Jones, she tells you of the need to maintain accumulated sales totals for each associate and asks you to add a field to contain this information in the database.

Step 1—Access the on-line Help feature of your database software program. Review the instructions for adding and **deleting a field** in an existing database.

Step 2—Remove the "Age" field from your database.

HELPFUL HINT

Review your software's on-line Help feature on how to add data to records once the structure of a database has been changed.

Step 3—Add a field named "Accumulated Sales". Enter the sales data that is shown below for each individual.

Accumulated Sales Figures

Associate's Name	Accumulated Sales
Marilyn Watson	$745,839.00
Mary Beth Wilson	$1,387,338.00
Larry Candler	$89,090.00
Catherine Ruth Jones	$6,347,333.00
Rick Schultz	$3,343,340.00
Marian Schmel	000.00

Step 4—Resort the database in ascending order.

Required Output: A printout of the revised database's contents using the default report format.

Other Activities and Questions

1. List at least five examples of databases that are used in the business world.

2. What fields would likely be contained in a database that relates to magazine subscribers?

3. Create a database that contains the names, addresses and phone numbers of your personal contacts.

Key Words and Phrases

ascending order—An alphabetic order that begins with words starting with the letter "a" and ends with words that start with the letter "z".

character—Alphabetic letters or special characters.

database—A collection of related records.

data type—The type of data contained in a field (alphabetic, numeric, date, currency, etc.).

default data entry format—A standard data entry form used by the software.

default format—The form in which data automatically appears.

default report format—A report format that is automatically produced by the software.

deleting a field—Removing a field from the database.

deleting a record—Removing a record from a database.

field—A group of related characters (i.e. first name, last name).

insert a new record or row—Adding a record to a database.

modifying the structure—Altering the basic definition of a database that generally refers to adding or removing a field.

record—A group of related fields.

sort records—A special arrangement of output from a database.

Intermediate Database Skills

Databases are designed to serve the special needs of a business. The information that databases store needs to be accessible to decision makers in the form they need it so that strategic and timely action can be taken.

Before a database can be used, however, its structure must be defined. Data must be input, verified and the system tested. Frequently, more than one table is created for a database and related to other tables if necessary. For example, a mail-order company might want to keep records of individual sales in a database. The database table which holds sales records for the same company, however, needn't contain data related to the business' inventory. The two tables can be linked by order numbers.

The hands-on activities in this unit provide the student with an opportunity to define a database that contains more than one table and to enter, sort, modify and search the database's contents.

Preview

When a database is defined, its purpose must be considered. The names of the fields contained in the database's records should be descriptive in nature and sufficient in number and type to hold the data and fulfill the database's purpose.

A database may consist of more than one table. Most computer software has the capacity to link more than one related table. This type of application is known as **relational database software.**

Relational database software uses a **key field** so that one table can be associated with another.

▶ ### Activity 1 Using Key Fields and Calculating Fields

In this activity you will:

- Create a **multitable database** that can be linked or associated with a **key field**
- Create a field in a database to be used for a **calculated field**

Situation:

Ms. Jones asks you to create a database that contains information on the owners of the rental units handled by Westwind. She intends for the database to be used by the employees for preparing income and maintenance reports, invoices as well as mailing

▼

activity 1 is continued on page 48

labels. Also, Ms. Jones asks that you create a database of the customers who have rented from Westwind Properties. She plans to use this information for marketing.

You decide that the database should be made up of two **tables** or files and that the unit number be the key field shared by both tables.

Step 1—Read the material in your database software's on-line Help that relates to defining a database and printing out its structure.

Step 2—Define a table or file that contains the fields shown in Table 1 - `Owner_Info`. Save the newly created database to disk. Be sure to format fields to contact dollar amounts as currency.

HELPFUL HINT

Make sure that the field sizes are large enough to hold the data. The data to be entered in this table is displayed later in this chapter.

Table 1 - Owner_Info

Number_of_Unit
Owners_Last_Name
Owners_Middle_Name
Owners_First_Name
Owners_Address
Owners_City
Owners_State
Owners_Zip
Owners_Phone_Number
Units_Phone_Number
Number_of_Bedrooms
Daily_Rate
Weekly_Rate
Monthly_Rate

Step 3—Define a table or file that contains the fields shown in Table 2 - `Rental_Record`. Save the database to disk.

Table 2 - Rental_Record

Unit_Number
Date
Number_of_Days
Number_of_Weeks
Number_of_Months

Rental_Rate
Renters_Last_Name
Renters_Middle_Name
Renters_First_Name
Renters_Address
Renters_City
Renters_State
Renters_Zip
Renters_Phone_Number

Step 4—Print out a copy of each table's **structure**.

Required Output: A printout of each table's structure.

Activity 2   Viewing Data and Correcting Errors

In this activity you will:

- View data stored in a database
- Find and correct errors in a database

Situation:

Ms. Jones supplies you with the raw data to put into the owner information table. Use your software's default data entry format to enter the information.

Step 1—Review the on-line Help material on how to enter data using the default data entry format. Enter the owner information that is shown below and on the following pages.

Unit 3234

Mildred L. Wheatman
819 Castlerock
Ft. Myers, FL 33942
Home Phone: 813-998-3433
Unit's Phone: 704-383-3387
Number of Bedrooms: 3
Daily Rate: $58.00
Weekly Rate: $350.00
Monthly Rate: $1,190.00

activity 2 is continued on page 50

Unit 3240

Billy R. Johnson
987 4th St. South
Ft. Myers, FL 33924
Home Phone: 813-654-4545
Unit's Phone: 704-433-5632
Number of Bedrooms: 3
Daily Rate: $60.00
Weekly Rate: $375.00
Monthly Rate: $1,400.00

Unit 3335

Marlene A. Roberts
384 Arcade Avenue
Clewiston, FL 33440
Home Phone: 813-485-6262
Unit's Phone: 704-586-4087
Number of Bedrooms: 2
Daily Rate: $58.00
Weekly Rate: $350.00
Monthly Rate: $1,190.00

Unit 3328

John D. Zemonick
18 Lilly Acres
Des Moines, IO 54832
Home Phone: 409-783-3465
Unit's Phone: 704-293-3386
Number of Bedrooms: 5
Daily Rate: $120.00
Weekly Rate: $735.00
Monthly Rate: $2,500.00

Unit 3239

John A. Scott
387 West Northlake Drive
Alexandria, VI 45874
Home Phone: 212-837-2333
Unit's Phone: 704-369-4422
Number of Bedrooms: 3
Daily Rate: $59.00
Weekly Rate: $385.00
Monthly Rate: $1,400.00

Unit 3339

Ms. Alice C. Brown
P.O. Box 144-A
Simpsonville, OR 87892
Home Phone: 419-883-5820
Unit's Phone: 704-586-3344
Number of Bedrooms: 2
Daily Rate: $57.00
Weekly Rate: $345.00
Monthly Rate: $1,173.00

Step 2—**Sort** the data contained in the owner information table by unit number in **ascending order.**

Step 3—**Scroll** through your table or database to determine if you have accurately entered the data. Correct any errors you discover.

Step 4—Using your software's **default report format,** print out the sorted data contained in the owner information table.

Required Output: A printout of the contents of the owner information table sorted by unit number in ascending order.

Activity 3   Renaming a Field

In this activity you will:

- Rename a database field

Situation:

You recognize that the individual records in the customer information table represent individual transactions. Ms. Jones told you that once you have entered the initial set of data it will be used on a daily basis.

Step 1—Using your system's default data entry format, enter the data shown below and on the following pages into the rental record table.

Unit 3234, for 3 weeks at $350.00 per week, starting 1/12/(current year)

Rented to: Mr. Alfred A. Jones
                90 Raindrop Drive
                Cincinnati, OH 37983
                Phone # 919-343-3444

activity 3 is continued on page 52

Unit 3335, for 2 weeks at $350.00 per week, starting 1/15/(current year)

    Rented to: Hector Honco
                      17 Palm Lane
                      Panama City, FL 33991
                      Phone # 904-338-3388

Unit 3328, for one month at $2,500.00 per mo, starting 1/17/(current year)

    Rented to: Robin Daley
                      144 Avenida Del Rio
                      Palm Dale, CA 07384
                      Phone # 323-383-8864

Unit 3239, for 1 week at $385.00 per week, starting 1/20/(current year)

    Rented to: Pat Waylon
                      Lark Lake Center, Apartment 2
                      Lark Lake, WI 98743
                      Phone # 313-849-4347

Unit 3339, for 2 months at $1,173.00.00 per month, starting 1/24/(current year)

    Rented to: Richard Simpson
                      878 Applegate
                      Douglas, GA 35478
                      Phone # 813-348-4484

Unit 3328, for 5 days at $120.00 per day, starting 2/25/(current year)

    Rented to: Alan Lawson
                      9919 4th Street North
                      Omaha, NE 23872
                      Phone # 919-884-3059

Unit 3339, for 3 weeks at $345.00 per week, starting 3/25/(current year)

    Rented to: James Ziegler
                      87780 Hightower Rd.
                      Provo, UT 87999
                      Phone # 212-838-8876

Unit 3335, for 1 week at $350.00 per week, starting 2/17/(current year)

    Rented to: Alexander McIntrye
                      104 Burgandy Lane
                      St. Louis, MO 98743
                      Phone # 345-876-4456

Unit 3239, for 1 month at $1,400.00 per month, starting 3/15/(current year)

    Rented to: Maxine Brantley
                    2420 Rossman Circle
                    F t. Denaud, ND 58201
                    Phone # 515-685-9034

Unit 3234 for 2 weeks at $350.00 per week, starting 3/22/(current year)

    Rented to: Delmar Bartlett
                    P.O. Box 9843
                    Van Wert, OH 44093
                    Phone # 414-384-3345

Unit 3240, for 3 days, at $60.00 per day, starting 3/08/(current year)

    Rented to: Mary Martin
                    2384 Tahiti Rd.
                    Naples, FL 33943
                    Phone # 813-842-4727

You determine that a field named "Base_Rate" would be more appropriate and descriptive than "Rental_Rate" in the rental record table.

Step 2—Review your software's on-line Help material on how to rename a database's field. Rename "Rental_Rate" to "Base_Rate".

Step 3—Print out a copy of the revised database's structure.

Step 4—Sort the individual records in the table in ascending order by date.

HELPFUL HINT

You might want to set up the page to print in landscape orientation.

Step 5—Print out a copy of the rental record table using the default report format.

Required Output: A printout of the rental record table structure showing the renamed field, and a printout of the rental record table, sorted in ascending order by date.

▶ Activity 4   Editing and Backing Up a Database

In this activity you will:

- Backup data using a database program
- **Edit** a particular record in a database
- **Add a record** to a database
- **Delete a record** from a database

activity 4 is continued on page 54

Situation:

Ms. Jones has just told you that Unit 3239 has been sold and that the new owner wants to live in it year round. Therefore, you need to delete Unit 3239 from the owner's table. Fortunately, on the same day that Westwind lost a client they gained one. So you need to add a record to the owner's table for a new property named Unit 4848.

Ms. Jones also tells you that the telephone number for Unit 3328 has been changed, so you'll need to **edit** the record. Because the information in the database is so valuable, you also decide to make a backup to disk.

Step 1—Review the material in your software's on-line Help feature that describes how to delete a record from a database. Delete record Unit 3239 from the owner table or file.

Step 2—Review the on-line Help material that relates to adding a record to a database. Add Unit 4848 to the owner's table. Use the information shown below.

> **Unit 4848**
>
> Peter L. Smith
> 9841 Westmoreland Drive
> Albany, GA 40489
> Home Phone: 404-323-0825
> Unit's Phone: 704-329-3333
> Number of Bedrooms: 4
> Daily Rate: $65.00
> Weekly Rate: $395.00
> Monthly Rate: $1,350.00

Step 3—Search for Unit 3328 in the owner table and change the unit's phone number to 704-879-3514.

Step 4—Sort the database by unit number in ascending order.

Step 5—Review your on-line Help feature on how to select specific fields within a table to print out. Print out a custom report that displays the unit number, the owner's name, the three rates and the owner's telephone number.

Step 6—Make a backup of the database to disk. If your software doesn't have a backup feature, use your operating system to make copies of the database and save it to disk.

Required Output: A printout of the modified database, sorted by unit number, that shows the unit's number, the owner's name, the three rates and unit telephone numbers, and a printout of the data disk's contents that demonstrates you have a backup of your database.

Activity 5 Querying a Database and Using Logical Operators

In this activity you will:

- **Query** an existing database
- Use **logical operators**

Situation:

Ms. Jones asks you to search or **query** your database to identify the unit that generated the most rental income and to print out a report that lists the owner's name, address, city, state and zip as well as the total amount of rental revenue that was generated.

Step 1—Review how to search or query a database using on-line Help and how to use logical operators.

Step 2—Review the on-line Help material of your software to determine if you can create a custom report that accumulates total revenue earned by each unit. Design such a report, if possible.

Step 3—Produce a custom report that includes the owner information for the rental unit that generated the most income.

Required Output: A printout of each unit's total revenue, and a printout that displays the unit number, owner's name, address, city, state and zip as well as the total amount of rental revenue generated by the unit with the largest sales.

Other Activities and Questions

1. Give an example of a database query that would be considered routine in the business world.
2. How does one determine the amount of storage space needed for a database?
3. What considerations do you feel enter into creating a database that would print out records on preprinted forms?
4. Create a database with multiple tables that contains student's names and a listing of courses.

Key Words and Phrases

add a record—Either inserting or appending a record to an existing database.

ascending order—A method of sequencing that proceeds from the beginning or least (the number "1" or the letter "A" and going to the last or largest (the letter "Z" or the largest number in the sequence).

calculated field—A field in a database that contains data that has been derived by mathematical operations.

default report format—The report format that is automatically supplied by the database software.

delete a record—Selectively removing a record from a table.

edit—Changing the contents of a particular field.

key field—A field which is common to all tables in a database.

logical operators—Words that are used in database queries that define the scope of the search or operation.

multitable database—A database that contains more than one related table in its structure.

query—Asking for retrieval of information from a database.

relational database software—Database software that is capable of relating data contained in more than one table.

scroll—Moving through the records in a database.

sorted—Records that are arranged in a particular order.

structure—A listing of the table's field names and characteristics.

table—A collection of related records.

table's structure—Listings of field names, size and type.

Advanced Database Skills

Producing special screen layouts and unique reports dramatically expand the useful work that can be done with database software. With customized data-entry screens, for example, people with limited computer experience can be prompted to enter data into a database. With a properly designed database application, special reports such as letters, mailing labels, or other documents can be created and incorporated into a database application.

Business computer users must know how to create databases, how to design special applications, and be able to modify an existing database as circumstances dictate. This lab unit will provide you with an opportunity to work with more advanced database tasks allowing you to become more familiar with typical workplace challenges associated with computers.

Preview

In the workplace, clerical, secretarial and other support staff are typically the individuals who enter data into a database. The database applications are usually developed by someone else.

Menus are created by the database's designer to prompt users to enter data or request actions. A selection is made and the action is either performed by the computer or the user is prompted by more menus to supply additional direction. Databases store individual records. These records can be edited, deleted or otherwise manipulated. Additional records can be added to the database and specialized reports produced.

Your value as an employee is tremendously enhanced if you know how to design an **interactive database application** for a business.

▶ Activity 1 Customizing a Data Entry Form

In this activity you will:

- Develop a **customized data entry form**

Situation:

Ms. Jones shares with you that she would like to create a customer database application that can be used for marketing purposes. She tells you that she wants the database to have menus that give users options and the ability to create specialized reports, such as a list of customers who have visited before and at what time of year.

▼

activity 1 is continued on page 58

Ms. Jones also wants a **customized data entry form** to make it easy for the office staff to use the database.

You decide to design an application that uses data contained in more than one table.

Step 1—Open the table `Rental_Record` that you created in Activity 1 of Intermediate Database Skills. Modify the table's structure by adding a new column named "Visited". This column will be used to indicate whether the individual has been a prior customer of Westwind Properties. Either the word "yes" or "no" is to be entered in this field.

Step 2—Read in your on-line Help on how to create a customized data entry screen. Create a new input screen or layout form. Match the content and use a similar format to the form that is shown below.

HELPFUL HINT

Each input area on the screen is associated with a field in the database.

```
                    Customer Information Form
                       Westwind Properties

    Unit Number:

    Customer Name:
         Last Name:       _____
         First Name:      _____
         Middle Initial:  _____

    Address:
    Street or Box:        _____

    City: _____  State: _____  Zip: _____

    Phone: (___) ___ _____       Past Visit? _____
                                                Yes or No
```

Step 3—Print out a copy of the input screen.

Step 4—Use the customized data entry screen to modify existing records in the `Rental_Record` table. Enter the word "yes" in the "Visited" field of each record in the database.

HELPFUL HINT

Existing records can be edited by using the data entry form. Review how to do this, if necessary.

Step 5—Print out a copy of one customer record showing that the new information has been added.

Required Output: A printout of the modified `Rental_Record` table's modified structure, a printout of the customized input screen and a hard copy printout of one customer's record showing the record has been changed.

Activity 2 Customizing a Report

In this activity you will:

- Develop a **customized report format**
- Create a database report that appears as though it was created as a word processing document

Situation:

Ms. Jones wants to send a marketing letter to customers who have previously rented from Westwind Properties. Ms. Jones would have had to go through each billing record from prior years to identify customers. Then, she would have had to distribute the stack of billing records among the office staff and have an original letter be prepared for each customer. The process was long and tedious.

You explain to Ms. Jones how she can use the existing customer database and the software's report generating feature to automate the entire process of sending prior customers a marketing letter.

Step 1—Read in your on-line Help how to design a customized report.

Step 2—Use the report feature of your database program to create the letter shown below. This is the letter to be sent to customers. Print out a copy of the letter.

Date:

First Name, Last Name
Address
City, State Zip

Dear (First Name):

We'd like to invite you to return to our area and stay in a **Westwind Properties** unit.

Book a vacation with us now and we'll provide you with a 10% discount.

Call us now and tell our reservationist you received this letter. We'll be thrilled to hear from you.

Sincerely,

Blaire Jones
Manager

---

activity 2 is continued on page 60

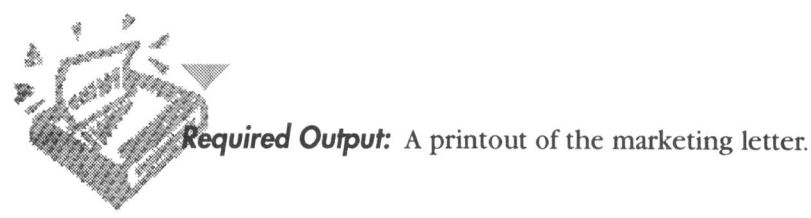

Required Output: A printout of the marketing letter.

Activity 3 Developing an Interactive Application

In this activity you will:

- Develop an interactive database application

Situation:

Ms. Jones is very interested in your progress on the customer database application. She wonders how easy it's going to be for the office staff to use. You explain that you are preparing a screen or menu that lets the user choose which function he or she wants to perform.

HELPFUL HINT

Writing the interactive application is far more complex than designing a screen for prompting users.

Step 1—Carefully review your on-line Help on how to develop an interactive database application. Also, read the sections explaining how to prepare a customized database application and how to **append** new records.

HELPFUL HINT

You must know how to append records to make option number 1 in the menu function properly.

Step 2—Create the screen shown below that prompts the user for input.

```
Westwind Properties Customer Information Database

1. Enter Customer Data
2. Merge Customer Database with Letter
3. Print Mailing Labels
4. Exit
```

Step 3—Print out a copy of the menu for Ms. Jones to review.

Step 4—Create your interactive application. Test each option of your application to make sure it works correctly.

HELPFUL HINT

Use logical operators to restrict your test for output to one record.

Print out a copy of the interactive database application's **script**.

Step 5—Merge the customer' names and addresses who have previously rented from Westwind Properties into the letter (customized report). Print out a copy of each letter.

Step 6—Print out a set of mailing labels (in actuality a second customized report) for each customer who receives a letter.

Required Output: A printout of the menu screen, a printout of the application's script or sequence of instructions, a printout of the marketing letter to send to each customer who previously stayed at Westwind Properties and a printout of the mailing labels for each marketing letter to be sent.

Activity 4 Importing, Querying and Exporting

In this activity you will:

- Import a file from a spreadsheet into a database
- Query a database utilizing multiple related tables
- Prepare a database file for export to a spreadsheet

Situation:

You realize that importing files from your spreadsheet into your database has the potential of saving time. You decide to conduct an experiment to see if you can do this.

Step 1—First, open the spreadsheet document **SSREV** that you prepared for Activity 1 in Advanced Spreadsheet Skills. This spreadsheet contains the names and numbers of the rental units owned by Westwind Properties as well as other data.

Create a new spreadsheet from the old one that contains just the rental units' numbers and their weekly rate. Save the new spreadsheet as **SSREV3**. Export the revised spreadsheet to a data file on a disk.

Step 2—Print out a copy of the exported file. It should contain the data shown below.

HELPFUL HINT

You should be able to produce the output using an operating system command.

Unit Number	Weekly Rate
1101	$375.00
1102	$420.00
1103	$385.00
1104	$415.00
1105	$435.00
1106	$450.00

activity 4 is continued on page 62

Unit Number	Weekly Rate
1107	$435.00
1108	$385.00
1109	$390.00
1110	$430.00
1111	$410.00

Step 3—Open your database software. Review your on-line Help on how to import a data file from another source.

Step 4—Next, import the spreadsheet file you just saved to disk into the owner's information named Table 1 - **Owner_Info**. This table was created as part of the Intermediate Database Skills activities.

Step 5—Query the database after importing the spreadsheet file to show the records which have only partial records.

HELPFUL HINT

One way to identify the database records that contain partial records would be to query the database to display those records that don't have a value in the owner field.

The records that are identified by the query should be the ones owned by Westwind Properties.

Step 6—Enter Westwind Properties as the owner of each partial record.

Step 7—Query the database so that it lists the owner's names for all rental units and the rates associated with the units. Store the results in a table. Print out the new table.

Step 8—Read in your on-line Help how to export a data file. Prepare the table containing the owners and weekly rates for exporting into a spreadsheet. Print out the contents of the data file.

Required Output: A hard copy printout of a spreadsheet data file that contains the names of the rental units owned by Westwind Properties and the weekly rates, a hard copy printout of the data contained in Table 1 - **Owner_Info** which shows the partial records for the rental units owned by Westwind Properties, and a hard copy printout of the exported files' content.

Other Activities and Questions

1. Can your database program perform math? If so, how is it accomplished?
2. Can a part of an interactive database application be designed to modify the existing database?
3. Can your database software be used to generate graphics? If so, how?
4. List at least five examples of how a database query might be used in the "real world".
5. Can more than one field or column be duplicated in related tables or can there only be one identical field?
6. Could you import custom designed graphics screens into your database software (i.e. from a paint program or a word processor)?
7. In what file format does your software export data?
8. Can you change the name of a table using your software? If so, how?
9. List at least three examples of custom designed reports.

Key Words and Phrases

append—The adding of records to an existing database.

customized data entry form—A data entry form which is designed with a unique title and labels.

customized report format—A report which contains information that has been specially arranged.

interactive database application—A program which has been written to prompt users to input data, perform manipulations and generate reports.

menu—A series of choices from which the user of an interactive database application may select the specific work they wish to accomplish.

script—A series of commands that cause an interactive database application to function.

Part IV
Presentation Graphics Skills

Beginning Presentation Graphics Skills

Many businesses today are saving money by creating their own graphics presentations. This is made possible by increasingly more powerful desktop computers and sophisticated software. Customized graphics images are being routinely produced in offices throughout America.

Input for graphics slide presentations can be obtained from digital cameras, video tape, the INTERNET, color scanners, dedicated drawing or painting programs, pages imported from software, such as spreadsheets, and with specialized presentation graphics software.

The output of the presentation graphics is varied. Once an image is produced, it can be output to specialized equipment that produces a 35mm slide, a color-laser printer, a device that produces a color photograph, on monitors on a company-wide **intranet**, a video tape, or a **WORM Drive**. The possibilities are limitless and even include the image being projected directly from the computer on to a color **LCD projector** for audiences to see.

There is little doubt that you will be faced with opportunities in the future to present ideas to others. You will be able to enhance the effectiveness of your communication by the skillful use of desktop presentation graphics software.

Preview

Presentation graphics software is used to design professional-quality screens or slides that can be output to paper, overhead transparencies, 35mm color slides or projected on to a large screen.

A number of common features are shared by most presentation graphics software applications. Text and images can be placed on the screen and then resized, colors can be changed and each graphics **object** can be moved. A variety of color schemes can be utilized to change the appearance of a slide and a group of screens may be combined to work as a slide show.

Activity 1 Creating a Presentation Graphics Slide

In this activity you will:

- Access the on-line Help feature of the presentation graphics software
- Place text on a presentation graphics slide
- Preview a slide in the process of being designed
- Spell check a presentation graphics slide
- Alter the size, style and font of text that appears on a previously saved slide
- Alter the color of text on a presentation graphics slide
- Relocate text on a slide

Situation:

Ms. Jones invites you into her office to discuss a **marketing plan** she is developing for Westwind Properties. Ultimately, she wants you to create a slide presentation that she can use in the office or take on the road.

She also requires that you first create two colorful examples of a **title slide** so that she has two samples from which to pick. Each slide should contain the name of the company and the motto:

Westwind Properties
"Time Away from It All"

Step 1—Open your presentation graphics software and prepare to create the first of the two title slides.

Step 2—Enter the first line of text, "Westwind Properties," in 18-point font size. Center the title both horizontally and vertically.

Step 3—Position the cursor below the first line of text. Center the motto, "Time Away from It All," using italics and a 14-point font size. Save the first slide under the name of `slide1`.

HELPFUL HINT

Review how to select text and graphic objects. Once you have selected an object you can move or resize it.

Step 4—Use your on-line Help feature to review the steps needed to move text on the screen.

Step 5—Next, without re-entering the data from the keyboard, move the words "Westwind Properties" to the top of the screen and increase its font size to 36-point. Be sure that the slogan *"Time Away From It All"* is still centered on the screen. Change its font size to 18-point.

Step 6—Change the color of the text on the slide.

Step 7—Spell check the slide. Make any necessary changes. Use the Save As feature of your software and save the second slide as **slide2**.

Step 8 - Use the software's **preview feature** to decide if you like the appearance of both **slide1** and **slide2**.

Required Output: Two presentation graphics files saved as **slide1** and **slide2**.

Activity 2 Using Clip Art

In this activity you will:

- Access the presentation graphic software's clip art library and insert an image on a slide
- Resize a clip art image on a previously saved presentation graphics slide
- Relocate the position of a clip art image on a previously saved presentation graphics slide

Situation:

Ms. Jones, upon reviewing **slide1** and **slide2**, decides that both need "some kind of art work." She encourages you to use your imagination and to experiment. She asks you to prepare a couple of examples for her to review.

HELPFUL HINT

Presentation graphics software usually contains a clip art library. Select the image you want and insert it into your slide.

Step 1—Open **slide1**. Access the **clip art library** of your software and select an image that you feel is appropriate. **Insert the image** on **slide1**. **Resize** and **reposition** the graphics image on the screen until you are satisfied. Relocate the text, if necessary to accommodate the graphic's image. Preview how the actual output would appear. Make any changes you feel are necessary to make the slide more attractive. Save the image under the name of **slide3**.

Step 2—Retrieve **slide2**. Select another clip art image. Place it on your slide. Resize and position it to your liking and save it under the name **slide4**.

Required Output: A file named **slide3** that contains the same text data as **slide1** and a clip art image, and a file named **slide4** that contains the same text data as **slide2** as well as a clip art image.

Activity 3 Modifying a Slide and Printing a Copy

In this activity you will:

- Copy and paste text on a slide
- Copy and paste a graphics image to a slide
- Alter the orientation of an existing presentation graphics slide
- Produce a hard copy output of a presentation graphics slide

Situation:

Ms. Jones asks you to produce a **hard copy printout** of a new slide that contains parts of previously saved slides (**slide3** and **slide4**).

HELPFUL HINT

Review the on-line Help feature of your software that explains how to produce a hard copy printout of a slide.

Ms. Jones decides that she likes the text on **slide3** and the clip art on **slide4**. She asks you to combine them so that a new slide contains the text of **slide3** with the clip art image that is displayed on **slide4**. Save the new slide as **slide5**.

Step 1—Open **slide3**. **Cut** or remove the text contained on **slide3**. Open a new screen. Place the text on the new screen. Save the resulting slide as **slide5**.

Step 2—Next, open the file named **slide4** and cut out the graphics image. Close **slide4** and open **slide5**. **Paste** the image contained in your **clipboard buffer** on to **slide5**.

Step 3—Resize and reposition the new screen so that it appears the way you want. Finally, spell check your graphics screen and make any necessary changes.

Step 4—Save the final screen as **slide5**. Produce two hard copies of the screen for Ms. Jones to review. One should be in a **portrait orientation** and the other in **landscape orientation.**

Required Output: Two hard copy printouts of **slide5**, one in landscape orientation and the other in portrait.

Other Activities and Questions

1. What is the size, in bytes, of the presentation graphics slides you prepared in Activity 1? Is there any difference between the two? Why or why not?

2. Design a title slide that contains your name and address.

3. Are there differences in the size of the files named **slide1** and **slide3**? If so, what is the difference and why?

4. How many clip art images are there in your presentation graphics software?

5. Add one or several clip art images to the file you created that contains your name and address.

6. Where is a graphics image or text that has been cut stored inside the computer?

7. Does the "printing" of a hard copy presentation graphics slide always produce a "paper copy"? If so, why? If not, why?

8. Can you print a larger image on a piece of paper using landscape orientation? If so, why? If not, why?

9. Select a graphics image from the clip art library and paste it on a slide. Produce a hard copy.

Key Words and Phrases

cut—The physical removal of an image or part of an image from the screen.

clip art library—A large number of graphics images that are part of many presentation graphics software packages.

clipboard buffer—Space in random access memory that is reserved for the storage of an object that has been cut from a document or screen.

hard copy printout—Output that is recorded on paper.

insert the image—Making a copy of a clip art image, transferring it to the presentation graphics screen that is being created, and attaching it.

intranet—An organization-wide computer network, similar to the World Wide Web in appearance, that many businesses are now adopting for internal communications.

landscape orientation—Output oriented in a horizontal manner.

LCD projector—**L**iquid **C**rystal **D**isplay is a technology that allows output from the computer to be generated on an otherwise transparent flat glass panel. The panel uses its own light source or may be placed on a standard overhead projector. The resulting image is projected onto a screen or the wall.

marketing plan—An organized effort to promote a business, its products or services.

object—A segment of text or graphics which can be manipulated as a unit.

paste—Placement of an object on a graphics image or document.

portrait orientation—Output that is arranged in a vertical fashion.

preview feature—A characteristic of software that allows you to see how the output would appear.

reposition—Moving the graphics image to another physical location on the same screen.

resize—Either shrinking or expanding a graphics image.

title slide—One of the first slides that appears in a presentation.

WORM drive—**W**rite **O**nce **R**ead **M**any times drives allow for the direct recording of digital data onto blank CDs.

Intermediate Presentation Graphics Skills

*M*any *businesses now have the software and computer power to produce their own graphics slide show. One of the primary reasons for creating a graphics slide show is to make presentations for marketing purposes.*

In this unit, your employer asks you to create a simple **slide show** *to introduce customers to your company and its services.*

Preview

Presentation graphics software generally has the capacity to alter text colors, background and foreground colors. Presentation graphics software, also, generally has a **drawing feature** associated with it as well as the capacity to use **clip art.**

Custom designed slides can be assembled in a sequence to tell a story. The designer of such a presentation graphics slide show can select the order in which any particular slide appears and the **transition** used to change from one slide to another.

▶ Activity 1   Creating a Master Slide

In this activity you will:

- Create a **master slide**
- Produce a simple graphic image on a slide using the presentation graphics drawing feature

Situation:

Ms. Jones tells you that she wants to give potential customers access to an impressive computer slide presentation that introduces Westwind Properties before the customer works directly with a salesperson.

She hasn't decided on the contents of the presentation yet, but does know that she wants you to design a company logo to be used along with the words "Westwind Properties" that is to appear on every slide. She asks you to start working on the master slide and tells you that she will get back to you later with the content of the individual slides.

Step 1—Open your presentation graphics software. If necessary, review the on-line Help materials on how to draw an image for use on a slide.

▼

activity 1 is continued on page 72

Step 2—Create a logo of your own design. Keep the drawing simple so that the detail is not lost when it is reduced in size. For example, use straight lines for mountains. Use a 12-point font size for the words "Westwind Properties."

HELPFUL HINT

The logo should be designed small so it can be placed in a corner (as the master slide is going to be used as a backdrop for every slide).

Step 3—Once you are satisfied with the master slide, save it to disk under the name `slide_m`.

Step 4—Produce a hard copy printout of the master slide.

Required Output: A hard copy printout of a master slide containing the logo of your own design and the words *"Westwind Properties."*

Activity 2   Creating a Properly Aligned Series of Slides

In this activity you will:

- Create a series of individual slides that together communicate a concept
- **Align text** on a presentation graphics slide

Situation:

Ms. Jones reviews the logo you designed and likes it. She has decided on the following content for each slide.

Step 1—Read the description for each of the slides that is shown below:

Slide One: Create a **title slide** containing the name of the company, Westwind Properties, in large letters. Retain the logo that is part of `slide_m` but remove the 12-point font size of the company name.

Slide Two: Feature the words "Lose Your Worries in the Mountains" and include a graphic image that is compatible with the message. Ms. Jones has told you that you can either create a drawing of your own or use clip art.

Slide Three: This slide should highlight the amenities of the region (the national forests, rivers, views and lakes). Make sure to align any text that you use on the side.

Slide Four: This slide should describe what is typically contained in Westwind Properties rental cabins (such as cable television, fully furnished, VCR, dishes, etc.)

Slide Five: Viewers should be invited, while spending their vacation in the area, to let Westwind Properties help them explore the possibility of purchasing a vacation home.

Slide Six: This slide should remind people that Westwind Properties is a full-service real estate company and people wishing to sell their property can list it with your company.

Slide Seven: This slide should tell the customers that the slide presentation has been completed and invite them to speak with one of the representatives in the office to discuss their property needs.

Step 2—Design each of the seven slides described above and save a copy of each to disk.

HELPFUL HINT

Remember that you will be using `slide_m` as the background for each of the slides you create. Use the Save As feature to save each new slide under a different name.

Step 3—Produce a hard copy printout of each slide for Ms. Jones' review.

HELPFUL HINT

Review the on-line Help materials on how to produce hard copy printouts of a slide show. Determine how to avoid producing full-sized page copies of each slide to conserve paper.

Required Output: A hard copy printout of each of the seven slides.

Activity 3 Modifying Text Objects and Creating a Slide Show

In this activity you will:

- Recolor a selected portion of an existing text object
- Combine a series of individual screens into a multi-screen slide show
- Manually change from one screen to another in a slide show

Situation:

Ms. Jones has reviewed the seven slides and has only one small change. That is, she wants you to use a different color for the text object contained on slide two. Now that she has approved the individual slides, you need to combine them into a slide show.

Step 1—Change the color of the text on slide two to a color of your choosing.

Step 2—Review your on-line Help materials on how to design a slide show.

HELPFUL HINT

Pay close attention on how to alter the ways that slides will transition from one to another.

activity 3 is continued on page 74

Step 3—Place the seven slides in order. Design the slide show so that changing from one slide to another requires **manual input** from the user. Vary the transitions used to change from one slide to another.

Step 4—Save your slide show under the name `hi_west`.

Required Output: A hard copy printout of the modified slide two, and a slide show, requiring manual input, saved to disk under the name `hi_west`.

Activity 4 Editing a Slide Show

In this activity you will:

- Edit an existing presentation graphics slide show

Situation:

Ms. Jones is very pleased with the slide show. She does ask, however, that you change the order of the slides. Specifically, she wants slide two to appear after slide five.

Step 1—Review your on-line Help materials on how to change the order of a slide presentation.

Step 2—Edit or change the order of the slides so that slide two appears after slide five in the new slide show.

Step 3—Save the slide show to disk under the name of `hi_west`.

Required Output: A hard copy printout of the slide show that indicates that the sequence in which the slides are presented has been altered.

Other Activities and Questions

1. Is it possible to direct output from a slide show to video tape? If so, how? If not, why?

2. Does your software have a feature for preparing a master slide template? If so, describe in writing how it works.

3. Prepare a slide show, consisting of less than a dozen slides, that describes your class to other students who are considering taking the course.

Key Words and Phrases

align text—Text which is even vertically.

clip art—Artwork that may be placed in slides.

drawing feature—Part of the presentation graphics software program that provides for creating artwork.

manual input—Advancing from one slide to another with input from either the keyboard or the mouse.

master slide—A type of slide template that is used to display the same information upon each slide.

slide show—A series of slides that are designed to be displayed in a specific sequence.

title slide—The introductory slide.

transition—The manner in which one slide is closed and another is opened to the viewer.

Advanced Presentation Graphics Skills

*P**resentation graphics software lets businesses create superior presentations while spending less money. This increases their marketing reach and their sales. Presentation graphics software has many features that help users create impressive slide shows. For example, users can import images from other applications, vary the default **color schemes**, create self-running presentations, and even solicit input from users so they can view different paths in the presentation.*

The series of activities in this unit provides you with opportunities to explore the upper limits of presentation graphics software.

Preview

One advanced feature associated with presentation graphics software lets you create a **bullet list** on a single slide, that allows you to **build one line at a time.** That way the pace of a presentation can be better controlled.

Also, most presentation graphics software allows you to import graphics images from a variety of sources. Therefore, you aren't limited to the clip art library that is included with the software. One available source of graphics images, for example, consists of three CD-ROM disks and contains over 30,000 images, any of which can be imported into a presentation graphics slide.

Most presentation graphics software provides for the creation of a slide show that allows for user input to direct which **logical branch** or path is followed. Slide shows can also be made that are self-running.

When you can include mastery of advanced presentation graphics skills in your list of specialized knowledge you will indeed be a valued employee.

▶ Activity 1   Creating Slide Show Special Effects

In this activity you will:

- Add text to a **graphics object** on an existing slide
- Create a bullet list slide upon which the individual items appear at different times
- Vary the time that individual slides are available for viewing in a multi-screen slide show
- Use various **transitions** in a presentation graphics slide show
- Create a slide show with a logical branch

Situation:

Ms. Jones received positive feedback on the **hi_west** slide show you prepared. However, the format of this slide show doesn't work as well when she makes business presentations in the community, so Ms. Jones would like you to create an alternate version to "take on the road." Specifically, she would like to change the master slide, and she wants to insert a slide that contains a bullet list that builds one line at a time.

HELPFUL HINT

Prior to beginning this hands-on exercise, make sure that the previously saved intermediate slide show, **hi_west**, is available.

Step 1—First change the master slide. Open **slide_m** that you created in a previous lab. Reduce the size of the text and move the text "Westwind Properties" to overlay on your graphics object. Save the altered slide.

Step 2—Read in your on-line Help how to build a bullet list. Create a slide with the following bullets that appear one at a time after input from the user.

- More than 135 rental units
- More than $5,000,000 in revenue
- More than 75 homes sold
- More than 1,000 satisfied customers

Step 3—Alter the default transitions between the slides in the **hi_west** presentation and vary the time that each slide is displayed.

Step 4—Save the new slide show as **hi_prs**.

Required Output: A printout of the revised **slide_m**, a printout of the bullet list slide with the instructions that show the bullets will build one line at a time, and a printout of the new slide show, **hi_prs**, with the instructions that shows varied slide transitions and viewing times are used.

Activity 2 Altering Slide Images and Importing Graphics

In this activity you will:

- Alter the background of an existing slide
- Alter the color scheme of an existing slide
- Alter the **shading** of an existing slide
- **Import a graphics image** and insert it on an existing slide

Situation:

Ms. Jones just told you that she has an extensive clip art library on her computer. She wants you to see if you can import a clip art image from the library and incorporate it

activity 2 is continued on page 78

into a slide. She is particularly interested in having an image that relates to wildlife or the outdoors.

Step 1—Review in your on-line Help how to import a graphics image. Your instructor may supply you with a graphics image to be imported into a new slide. Your GUI's paint program might be a source for graphics images as well.

Step 2—Identify the image you are going to use and import it into the slide. Print out a copy of the slide.

Step 3—You decide the slide would look better with a different background, color scheme, and shading. Experiment with different combinations until you find one you like and print it out.

Required Output: A printout of the first version of the slide and a printout of the second version of the slide.

Activity 3   Creating a Runtime Version Slide Show

In this activity you will:

- Insert and incorporate a new slide into a previously saved slide show
- Create a slide show with interactive logical branches
- Produce a **runtime version** of a slide show

Situation:

Ms. Jones also wants to send a third version of the slide show to prospective customers who have compatible computers. Ms. Jones wants to include **digitized photographs** of rental units and houses for sale in the slide show. She wants the customers to be able to choose whether they want to look at the rental units or the houses.

In order for the presentation to be run on other computers, you realized you'll need to create a runtime version of the slide show. You'll also need to include a logical branch that lets customers view information on either the rental units or the houses.

Step 1—Read in your on-line Help how to create a branch in your slide show and how to create a runtime or automatic version of the presentation.

Step 2—Create six blank slides that just contain titles for House One, House Two, and so on. Ms. Jones will insert the digitized photograph files later.

Step 3—Revise the `hi_west` presentation so it includes a logical branch and **prompts** the customers, asking them whether they want to view the rental slides or the house slides.

Step 4—Alter the slide show so that it will run automatically instead of manually. Save the revised slide show as `hi_mail`.

Required Output: A printout of the `hi_mail` slide show that includes the instructions which will cause it to run automatically and be capable of branching.

Other Activities and Questions

1. How many different transition methods are available in your presentation graphics software?
2. How many different graphics file formats are there available for use? How are they distinguished from one another?
3. Are you programming when you create a slide show with logical branching? If so, why? If not, why?
4. What type of problems would you anticipate with sending runtime versions of presentation graphics slide shows to customers?
5. Obtain as many different graphics file types as you can and attempt to import them into a slide.

Key Words and Phrases

build one line at a time—A line of text that appears, individually, as part of a series of text lines on a slide.

bullet list—A series of related text lines on a slide that are preceded by a special character.

color scheme—A default color mixture usually provided in presentation graphics software.

digitized photograph—A picture, taken by a camera, that stores the image in a series of 1's and 0's that can later be manipulated by a computer.

graphics object—A digital image that can be manipulated.

import a graphics image—A likeness that originates from an outside source.

logical branch—A point in a presentation graphics slide show where there is more than one option for the user to view.

prompts—Solicit actions from the user; these appear on a screen.

shading—Any alteration in the pattern or intensity of a graphic's background.

runtime version—A version of a slide show that can run automatically without the support of the presentation graphics program.

transitions—The method by which one graphics slide image dissolves and the next one appears.

Part V
Operating System Skills

Beginning Operating System Skills

You need to understand a computer's **operating system** in order to take full advantage of its power. Otherwise, you will be dependent upon someone else to set up your system and determine how you work. You would have to get help every time you use your computer for something different.

A computer's operating system specifies the rules by which the hardware and software work together to accomplish results. There are commands that help you obtain and direct output, and programs that allow you to create system and data disks for backup and additional storage.

Operating system commands let you exercise control over your computing environment as well as to perform routine file maintenance, such as viewing a directory's contents, changing file names, or switching default drives. In this lab unit, you'll practice performing these basic operating functions.

Preview

An operating system performs two main functions. First, the operating system governs how a computer system's resources are managed and allocated. Second, the operating system provides an interface for interaction with the user.

A user can issue commands, via the operating system, that "tells" the computer to perform certain tasks that are necessary for work to be accomplished.

Activity 1 Understanding Disk, Directory and Screen Operations

In this activity you will:

- **Format a system disk** that contains a unique name
- Format a **data disk** that contains a unique name
- Produce a hard copy printout of a directory's contents
- Demonstrate **switching the default directory**
- Clear the display screen of its contents

Situation:

You report to work on your first day at Westwind Properties and are introduced to the other employees. Ms. Jones then takes you to your office where you see your desk, a computer, and a box of unopened disks and a printer.

Ms. Jones tells you that she purchased the computer and printer "from a friend" and she is not sure what software came with the computer and asks you to find out.

Ms. Jones also asks you to spend the next couple of hours becoming familiar with your office and the computer. You decide that one of the first things that you must do is turn on your computer and create a **system or start-up disk** to protect against hard drive failure.

Step 1—Turn on the microcomputer. After it has completed **booting**, examine the contents of the **default directory** by issuing the appropriate command to display the hard drive's contents on the screen.

Step 2—**Toggle** the printer "on" (or use the print screen command) so that the next command you issue will be printed. Issue the command, once again, that displays the contents of the directory. Toggle the printer "off", if necessary.

Step 3—Issue the command that clears the contents of the screen.

••
HELPFUL HINT

A system disk contains the data needed to boot a computer whereas a formatted data disk is only for data storage.
••

Step 4—Change the default directory, if necessary, to the one which contains the **utility** or program that allows for the formatting of a system or start-up disk and a data disk.

Step 5—Issue the appropriate commands to format a system or start up disk. Follow the prompts and insert the disk into the appropriate drive.

Step 6—Next, format a data disk.

Required Output: A hard copy printout of the default directory's contents, a formatted system disk and a formatted data disk.

Activity 2 Using Essential File Operations

In this activity you will:

- Display the contents of a directory
- Assess a file's characteristics
- Switch the default drive
- Change the name of a file

Situation:

You want to produce a hard copy printout of the names of any files that are contained on the system or start-up disk that you just created. You plan to maintain a copy of every disk's contents so that you can review them at your leisure and away from your desk.

Since you plan on making a **back-up** of your computer files to store at home, you decide to practice renaming a file to a less descriptive title.

Step 1—Insert the system disk in the drive. Change the default drive and issue the appropriate command to display its contents.

Step 2—Toggle the printer or issue the appropriate command to produce a printout of the disk's contents.

Step 3—Select a single file on the disk. Write down the original file name for future reference. Change the name of the file you selected to `altered`.

HELPFUL HINT

Be certain to keep a record of any file names that you might change.

Step 4—Toggle the printer "on" or issue the appropriate command to obtain a printout of the directory's contents. Toggle your printer "off", if necessary, and save the output.

Step 5—Rename the file on your system disk named `altered` back to its original name.

Required Output: A printout of the system disk's contents and a printout of the system disk's contents that shows the file you altered.

Activity 3 Accessing On-line Help

In this activity you will:

- Access and use the operating system's on-line Help feature

Situation:

Ms. Jones told you that she would like for you to be responsible for "teaching the non-computer-literate sales associates" how to use computers. You decide that one of the

activity 3 is continued on page 84

best ways to begin work on the project is to examine the possibility of using printouts of your computer system's on-line Help feature.

Step 1—Access the on-line Help feature of your computer system. Search the Help feature for information on how to format a system or start-up disk and a data disk.

Step 2—Read the information you discover. Issue the appropriate command to obtain a printout of the on-line Help information.

Required Output: A printout of information contained on your system's on-line Help feature that relates to formatting a system or start-up disk and a data disk.

Other Activities and Questions

1. What information, other than a file's name, is contained on the printout of the default directory's contents? How would the additional information be used?

2. Why would it be important to have a "bootable" system disk?

3. Would it be desirable to print out all of the Help files that are contained on the software?

4. Why would it be useful to be able to rename a file?

Key Words and Phrases

back-up—An exact copy of a file, directory or disk's contents.

booting—Turning on the computer's power switch.

data disk—A diskette created specifically to hold data.

default directory—The directory to which the computer is oriented at any given time.

directory—A unique storage area for related files.

formatting a system disk—Creating a "bootable" diskette upon which the operating system's key files are stored.

operating system—The rules by which computer hardware and software perform their tasks.

switching the default drive—An overt command issued to redirect input or output to a different storage device.

system or start-up disk—A disk which is capable of booting a microcomputer system.

toggle—Entering a particular key sequence so that an output device is activated.

utility—A program which performs a routine computing operation.

Intermediate Operating System Skills

*Keeping your files in order and protected is a necessity. One method of organizing files is through the use of directories or **folders**. You can, for example, store all related files in a directory or special storage area that you create. Once this is accomplished, you would know where important information relating to a particular group of files is located.*

Protecting files is also an important consideration. Computer systems could "crash" and you could lose your files. There are a number of ways to accomplish the task of backing up files. One obvious method is to make a copy of the files from the hard drive to disks or other storage media.

In this lab unit you have an opportunity to better organize files and make copies of critical information.

Preview

Knowing how to copy files from one disk drive to another or from one location to another is important. Frequently, in the employment setting, you find the need to replace a damaged file, back-up files or to provide other individuals with copies of crucial data.

You should know how to create new space in which to store files or groups of files. Managing crucial files is an essential skill. Part of file management includes being able to delete selected files.

Another set of knowledge that is useful is adjusting the time and date of a system. Unless your computer is using software that can adjust for a change to or from daylight savings time to standard time, you must do it yourself. Most computer systems provide internal calendars that track dates well into the 21st century. However, a computer's date may have to be changed if you, for example, crossed the international date line.

▶ Activity 1   Copying a File

In this activity you will:

- Copy a file from the hard drive to the data disk

Situation:

You decide to back up critical files contained on your computer in the office at Westwind Properties. One of the first files that you want to preserve is the file that is used to format data disks.

Step 1—Turn on your computer and insert a previously formatted or initialized disk into the disk drive.

Step 2—Find the operating system file, contained on the hard drive, that is used to format disks. Copy the file from the hard drive to the data disk.

Step 3—Produce a hard copy printout of the data disk's contents.

Required Output: A printout of the contents of the data disk which should contain the file that is used to format a disk.

Activity 2   Creating Specialized Storage Space

In this activity you will:

- Create specialized storage space on the hard drive and data disk
- Copy a file from one specialized storage area to another

Situation:

You decide to create a new storage area, named **MINE**, on the hard drive. This space will be used to keep a variety of important files for Westwind Properties. By having all of the files in one storage location you can quickly locate a critical file and utilize it to replace or repair a damaged file.

Step 1—Create a new **sub-directory** or storage space on the hard drive named **MINE**.

Step 2—Copy the file on the hard drive that allows you to format a disk into the newly created storage space on your hard drive named **MINE**.

Step 3—Change the default drive to the disk drive. Create a new sub-directory or storage space on the disk named **MINE2**.

Step 4—Copy the file you stored in **MINE** on the hard drive to **MINE2** on the disk drive.

Step 5—Produce a printout of the contents of **MINE** on the hard drive and **MINE2** on the data disk.

Required Output: A printout of the contents of **MINE** and a printout of the contents of **MINE2**.

▶ **Activity 3 Working with a Group of Files**

In this activity you will:

- Copy a **group of files** from one directory on the hard drive to another
- Copy a group of files from the hard drive to a data disk

Situation:

Ms. Jones tells you that she feels it is important for you to plan on backing up certain groups of files that are crucial to the successful operations of Westwind Properties. One group of files that she describes are the **accounts receivable files** which contain information on customers who owe the company money.

The accounts receivable files all begin with the letters AR and are followed by a number. The numbers correlate to the month of the year. For example, AR3 is the name of the accounts receivable file for the month of March.

Step 1—Choose a group of files (or your instructor will specify a group) from the main directory on your hard drive and copy the selected group of files to the newly created storage space named **MINE** on the hard drive.

Step 2—Copy the same group of files on the hard drive to the storage space on the data disk named **MINE2**.

Step 3—Produce a hard copy printout of the contents of **MINE** on the hard drive and **MINE2** on the data disk which show that the selected group of files have been successfully copied.

Required Output: A printout of the contents of **MINE** and a printout of the contents of **MINE2** on the data disk.

▶ **Activity 4 Deleting Files and Specialized Storage**

In this activity you will:

- Delete selected files from a directory
- Remove a directory

Situation:

Ms. Jones informs you that she wants all critical files that are maintained in the office to be contained in a directory or storage space named **CRUCIAL**. You realize that you will have to remove or delete the storage space **MINE** on the hard drive and **MINE2** on the disk.

HELPFUL HINT

Review how to remove directories or specialized storage space from a disk before attempting the steps below.

Step 1—Access **MINE** on the hard drive. Issue the commands that are necessary to delete all of the files.

Step 2—Access **MINE2** on the disk drive. Issue the commands that are necessary to delete all of the files.

Step 3—Produce a hard copy printout of **MINE** and **MINE2** that shows that the files have been deleted.

Step 4—Remove both **MINE** from the hard drive and **MINE2** from the data disk.

Step 5—Produce a hard copy printout of the hard drive's contents and the data disk's contents that show **MINE** and **MINE2** have been removed.

Required Output: A hard copy printout of **MINE** and **MINE2** that show each directory is empty and a printout of the hard drive and data diskette that show **MINE** and **MINE2** have been removed.

Activity 5 Adjusting the Time and Date

In this activity you will:

- Use the operating system to adjust the time and date of the computer system

Situation:

You were late for lunch because the time on your computer is wrong, so you want to correct it. After looking closely, you realize that the date is also incorrect.

Step 1—Use the operating system's command to change the date being displayed on your computer to December 1 of the current year and the time to be 3:00 PM.

Step 2—Print out a page that contains the revised time and date.

Step 3—Change the December 1 date and 3:00 PM time back to the correct date and time. Print out the correct date and time.

Required Output: A printout showing a December 1 date of the current year with a time of 3:00 PM. and a printout showing the date and time of the system have been changed back to the current date and time.

Other Activities and Questions

1. Name and describe at least three different methods that can be used to back up files.

2. Why would a user want to create sub-directories?

3. Describe a situation in which you would copy a single file or group of files from one directory on the hard drive to another.

4. Would Ms. Jones be able to take home disks containing stored files and place them on her computer? If yes, why? If no, why not?

Key Words and Phrases

accounts receivable—An accounting term which refers to the money which is owed to a business organization.

folder—A term, usually associated with a graphical user interface, that refers to storage space where related files are placed.

group of files—Files that are related to one another by their name or presence in a directory.

sub-directory—A directory within a directory.

Advanced Operating System Skills

A computer is configured to work in a particular "environment" (such as with a printer, touch screen, on a local area network, or with multiple hard drives). The operating system is used to define how the computer is to function. The operating system, thus, determines how a computer's hardware and software resources are utilized.

To effectively use a computer the user should know how to alter its **system configuration** so that it functions more efficiently. On occasion, **installation software** makes the necessary changes and at other times alterations must be specified manually.

Knowing how to restore and back up critical files as well as how to change file attributes, is also crucial to successfully using computers. In this lab unit, you will have an opportunity to work with each of these tasks.

Preview

A computer's operating system contains a number of special files. When a new device (such as a light pen) is installed, the file which orients the computer to the hardware that is attached must be changed. The names of files needed to make the light pen, for example, work properly are known as **drivers** and they must be identified.

Many specialized files can be altered directly by the user. Usually these are **ASCII** or text files. They may be changed by using a word processor.

A number of specialized **utilities** or "housekeeping software routines" are associated with an operating system. Some utilities can be used to change the way in which data is displayed and others perform useful functions such as changing a file attribute to **read only** or providing for password protection.

Circumstances in the business world determine how a computer is used. The operating system provides the user with tools needed to customize the computer's system configuration so that the computer can function effectively in the manner that is needed.

▶ Activity 1   Backing Up and Restoring Files

In this activity you will:

- **Back up** selected files
- **Restore** selected files that have been backed up

Situation:

Ms. Jones tells you that she wants to learn how to back up her files and restore them in the event the original file or disk is damaged. She also would like you to teach a class on this topic for all employees and to prepare a handout that can be used for future reference.

Step 1—Access the on-line Help materials that relate to backing up and restoring files. Read each section thoroughly.

Step 2—Utilizing your own words, write a draft of a step-by-step guide on how to back up and restore files. The guide is to be used as a handout for the class you teach to employees in your company, and for reference.

Step 3—Test your instructions thoroughly by backing up a file, then deleting and restoring it. Make any necessary changes in your instructions and print out the final copy.

Required Output: A step-by-step guide on how to back up and restore selected files.

▶ Activity 2   Modifying a File's Contents

In this activity you will:

- Alter the contents of an ASCII file
- Modify a key operating system file

Situation:

Ms. Jones asks you to change the contents of an ASCII file so that the company's name, Westwind Properties, appears on all of the company's computer screens upon turning on the power.

Step 1—Identify the file (typically an ASCII file) that needs to be modified to make specific text appear on the screen upon booting. You may need to refer to an outside text, or your instructor may provide you with the information you need.

Step 2—After you have identified the file, use your system's text editor or word processor to make the necessary changes in the file.

HELPFUL HINT

Save the file you modify under a different name so that you don't alter the manner in which your computer functions.

Required Output: A printout of the file that causes the company name, Westwind Properties, to be displayed upon booting.

Activity 3 Sorting Files

In this activity you will:

- Sort files in a directory

Situation:

Ms. Jones tells you that the directories on her computer's hard drive are not listed in any special order. You volunteer to show her how to sort the contents of a directory.

Step 1—Print out a hard copy of the unsorted directories.

Step 2—Read how to sort the contents of the hard drive using the on-line Help materials.

Step 3—Sort the directories by date.

Step 4—Print out a copy of the newly sorted directories.

Required Output: A hard copy printout of both the unsorted directory and the sorted directory.

Activity 4 Altering the System's Configuration

In this activity you will:

- Alter the microcomputer's system configuration

Situation:

After reviewing the guide that you prepared for backing up and restoring files, Ms. Jones asks you to help her install a scanner. You read the instructions that accompany the scanner and notice that the manner in which your computer operates needs to be altered.

Step 1—Retrieve the file that controls your system's configuration and print out a copy. Your instructor will tell you the name of the file and what specifically needs to be changed.

activity 4 is continued on page 94

Step 2—Use the text editor or word processor to make the necessary changes in the file. Save the file and print out a copy.

HELPFUL HINT

Save the file you modify under a different name so that you don't alter the manner in which your computer functions.

Required Output: A printout of the original configuration file, and a printout of the modified configuration file.

Activity 5 Changing File Attributes

In this activity you will:

- Alter individual **file attributes**

Situation:

Ms. Jones is worried that certain critical files might be accidentally erased. She knows that the operating system can be used for file protection but doesn't know how to do it. She asks you to prepare a brief outline for her on how to protect files from being deleted.

Step 1—Review the materials contained in your on-line Help on how to change file attributes.

Step 2—Select a file or your instructor will specify one. Change the file's attributes so that it is **read only** and can't be erased.

Step 3—Try to delete the file. Print out the error message indicating that the command failed to execute.

Step 4—Prepare a brief report for Ms. Jones on file attributes and how to change them.

HELPFUL HINT

There is more than one file attribute that can be changed. Include each of them in your instructions.

Required Output: A printout that shows an attempt was made to delete the protected file and a printout of the brief report for Ms. Jones that describes file attributes and how to change them.

Other Activities and Questions

1. How many different methods of backing up and restoring files can you identify?
2. What method of backing up files would you choose and why?
3. What are the various file attributes that can be changed?
4. How many ways can you sort files in a directory in your system?

Key Words and Phrases

ASCII file—A file that is stored in the format known as the American Standard Code for Information Interchange.

back up—A phrase used to describe making copies of crucial files.

drivers—Programs which make it possible for hardware to operate within a particular computer's environment.

file attributes—Those characteristics associated with files that determine whether specific operations can be performed upon them (i.e. deleted or altered).

installation software—Software which directs the installation of an application on a computer.

read only—The characteristic of a file that allows it to only be read rather than added to or deleted.

restoring—Replacing a file that has been deleted or lost.

system configuration—The characteristics assigned to a computer system so that its resources are allocated in a particular manner.

utilities—Programs which perform special housekeeping routines, such as file sorting.

Part VI
Graphical User Interface Skills

Beginning Graphical User Interface Skills

Graphical User Interface (GUI) technology is becoming the method of choice for working with computers. Earlier computer systems used a text-based or command line prompt. GUI is a visually based method of interacting with the computer that requires fewer keystrokes and makes computers easier for novices to use.

In a small business, very few individuals are totally proficient with a computer. Usually, there is one person who knows more than others. This individual then ends up with the responsibility of teaching other employees. In this lab unit, you must assume the role of this individual and help others learn how to use a computer's GUI.

Preview

A **graphical user interface** utilizes icons or symbols to represent the various actions that a computer can take. Users of a graphical user interface generally point a **mouse** or pen to select the application that they want to start and depress or **click** a button.

All standard operating system functions are included in a GUI. Knowing where the application or utility resides that does the work you want to do is an important consideration with a GUI.

▶ Activity 1   Activating and Using the GUI

In this activity you will:

- Initiate your computer's graphical user interface
- Use a **mouse** to make **menu selections** and to click and **drag**

activity 1 is continued on page 98

Situation:

Ms. Jones has a graphical user interface installed on her machine at work and home. However, she has continued to use the older, text-based operating system. She asks you to show her how to use the graphical user interface. You noticed that the same graphical user interface is installed on your computer.

Ms. Jones asks to prepare a list of instructions on how certain components of a graphical user interface are operated.

Step 1—Prepare instructions for Ms. Jones on how to use the GUI on her computer. Use either the **text editor** associated with your system's graphical user interface, a word processor or pen and paper to record the steps. Be sure to include basic instructions on how to properly use the mouse (i.e. "clicking" and "dragging").

Ms. Jones plans to give the instructions you prepare to the rest of the staff.

Required Output: A list of instructions explaining how to use your computer system's graphical user interface.

Activity 2 Opening and Closing GUI Applications

In this activity you will:

- Open and close an application program from within the graphical user interface

Situation:

After Ms. Jones mastered the steps necessary to start the computer system's graphical user interface, she asks you to show her how to "do some real work on the computer." Specifically, Ms. Jones wants you to show her how to open and close an application software package.

Step 1—Because Ms. Jones has already indicated that she wants everyone in the office to become more computer literate, you decide to write down the steps that are needed to open and close an **application.** Use the text editor associated with your graphical user interface to record the instructions.

Required Output: A list of instructions explaining how to open and close an application from within the graphical user interface.

Activity 3 Becoming Familiar with the GUI

In this activity you will:

- Identify the various parts of the GUI screen

Situation:

You take the initiative, based upon Ms. Jones' questions, and decide to create an illustration that identifies all of the major parts of the GUI's screen. When you explain your plan to Ms. Jones, she is enthusiastic and tells you that she would definitely like to use your guide for the office staff.

Step 1—Prepare a rough drawing of your computer system's default GUI screen and label the parts. Use the text editor associated with your GUI, a word processor or a pen and paper to explain the purpose of each component.

Required Output: A drawing of the default GUI screen including a brief narrative that explains each component's function.

Activity 4 Using the GUI's On-line Help

In this activity you will:

- Use the on-line Help feature associated with the graphical user interface

Situation:

You decide to prepare an orientation on how to use the GUI's on-line Help feature. You plan to explain in the instructions how to search for assistance on a particular topic and obtain a printout of what is contained in the Help file.

Step 1—Prepare step-by-step instructions on how to use your GUI's on-line Help feature. Use the text editor associated with your graphical user interface, a word processor, or pen and paper to prepare the instructions. Include how to search for help on a particular topic and how to obtain a printout of the on-line Help text.

Required Output: An original document which orients new users on how to use the on-line Help feature of the GUI.

Other Activities and Questions

1. Can font sizes be changed using the text editor associated with the GUI?
2. Describe the various programs or utilities that are included in your GUI.
3. Select a useful program (i.e. a simple paint program) or utility (i.e. hardware control utilities) associated with the GUI. Review its features. Write a set of instructions on how to use the selected program or utility.

Key Words and Phrases

application—A program represented by an icon on a Graphic User Interface screen.

click—Depressing a button on a mouse.

drag—Depressing a mouse button for a long period of time to either highlight text, expand boundaries or move an object.

GUI—Graphical User Interface.

graphical user interface—A group of graphic icons that work with or as part of an operating system that allow the user to issue commands to the computer.

menu selections—The specific selection of an option from a graphical user interface.

mouse—An input device which provides for pointing, clicking and dragging.

text editor—A simple text processing application.

Intermediate Graphical User Interface Skills

Graphical User Interfaces are becoming the standard with desktop computers of all types. Many individuals, however, lack the experience and the knowledge needed to work with a GUI.

There are a variety of routine skills that a user must possess to make effective use of a graphical user interface. For example, being able to open, re-size and close windows is important. Knowing how to interact with **dialog boxes**, use scroll bars, as well as how to change the look of the desktop are all important sets of knowledge to possess.

Once again, because you have knowledge of how to use a Graphical User Interface and you are the computer specialist at Westwind Properties, you are asked to help your employer and co-workers.

Preview

The programs that a computer is capable of running are represented by **icons** in a graphical user interface. Users select the icon they want and the program is started.

Programs that run under a graphical user interface share basic characteristics. Most have **scroll bars** that allow users to browse areas beyond the default display. Most GUI windows can be moved or resized.

You need to master the features of the graphical user interface on your computer to be successful on the job.

▶ Activity 1 Manipulating Windows

In this activity you will:

- Open and close windows
- **Resize** a window (maximize and minimize)
- **Reposition** a window
- Switch from one GUI window to another

activity 1 is continued on page 102

Situation:

Ms. Jones was very impressed with your GUI orientation and wants you to prepare a list of instructions on how to start programs or applications and "work with all of those windows."

Step 1—Review your computer's on-line Help feature on:

a. How to open and close a window or a program

b. Resizing a window

c. Repositioning a window

d. Switching from one window to another

Step 2—Create step-by-step instructions, in your own words, for each of the tasks described in *Step 1*.

Step 3—Test the accuracy of the instructions you prepared by performing each task on your computer.

Required Output: A hard copy printout of the instructions for each of the tasks.

Activity 2 Formatting a Disk

In this activity you will:

- Format a disk using the GUI

Situation:

Ms. Jones asks you to prepare a list of instructions for her on how to format a disk using the GUI.

Step 1—Access your GUI's on-line Help feature. Review the section that describes how to format a disk.

Step 2—Prepare a list of instructions, in your own words, on how to format a disk using the GUI.

Step 3—Test the accuracy of the instructions you prepared by actually formatting a disk using the GUI.

Required Output: A printout of the instructions on how to format a disk, and a disk that has been formatted using the system's graphical user interface.

▶ Activity 3   Using the Clipboard

In this activity you will:

- Utilize the GUI's **clipboard**

Situation:

You decide that it would be helpful to Ms. Jones if she understood how to use the GUI's clipboard. Therefore, you take the initiative and proceed to prepare instructions for her.

Step 1—Review the GUI's on-line Help feature on how to use the clipboard.

Step 2—Prepare a list of instructions, using your own words, on how to use of the GUI's clipboard.

Step 3—Test the clarity and accuracy of your instructions by utilizing the instructions you prepared on how to use the GUI's clipboard.

Required Output: A printout of the instructions on how to use the GUI's clipboard.

▶ Activity 4   Using a GUI Accessory

In this activity you will:

- Use the calendar program that is associated with the GUI

HELPFUL HINT

Some GUI's have a calendar feature associated with them. Other computers running a GUI may have programs with calendars associated with them. Determine which circumstance applies to you.

Situation:

Ms. Jones noticed that her computer has a calendar icon associated with it and asks you to help her get started using it.

Step 1—Review the on-line Help feature that relates to the calendar program and how to use it.

Step 2—Prepare a list of instructions for Ms. Jones, in your own words, on how to use the calendar program.

Step 3—Open the calendar program. Test the accuracy of your instructions by preparing an illustration for Ms. Jones using the data on the following page.

▼

activity 4 is continued on page 104

9:00 AM	Appointment with sales staff
10:00 AM	Meeting at Southeast Bank with Mr. Wilson
12:00 PM	Rotary Club meeting
1:30 PM	Begin interviews with secretarial candidates
3:30 PM	Take James to Little League practice
8:00 PM	Parent Advisory Committee Meeting

Required Output: A printout of the instructions on how to use the calendar program, and a printout of the sample day's activities.

Other Activities and Questions

1. Can you place graphics in the GUI's clipboard?

2. How does the Cut feature differ from the Copy feature? Why would you use one as opposed to the other?

3. How would your GUI's calendar handle the advance from 11:59 PM on December 31, 1999?

4. What features would you like to see on your GUI that aren't presently available? Why?

5. Pick a GUI component (i.e. the paint program, the clock, etc.) and explore *all* of its features.

Key Words and Phrases

clipboard—Storage space, usually in random accessory memory, where text or graphic objects that were cut or copied from a document may be kept until the need arises to paste them.

dialog box—A text and graphics prompt, usually associated with a Graphical User Interface, that receives input from the user so that specific action can be executed.

icons—A graphics object, displayed on a screen, that when selected causes a program to execute.

reposition—To move or relocate text, a graphic or a window.

resize—A method of changing the size of a window or object.

scroll bars—An area on the screen of a GUI program that allows for browsing.

Advanced Graphical User Interface Skills

You can alter the features associated with most GUI's to suit your individual preferences. You can change the color scheme, the arrangement of icons, default views and even the location to which output is directed.

A user would be wise to learn how to change the appearance of a GUI.

These lab units give the student an opportunity to demonstrate a understanding of how to change the GUI's default values.

Preview

Most GUI's can be tailor-made to meet the needs of individual users. For example, a GUI's screen saver (which is an animated graphics routine) that initiates after a period of inactivity can be changed as well as the time period that must pass before the screen saver begins.

GUI's have a default appearance when they start up. The default settings may have been determined by the software publisher, the individuals who manufactured the computer, the people who originally set up the GUI or a previous user. Regardless, the appearance of the GUI is subject to be changed.

Tool bars, icons, menus, and virtually every other component of a GUI can be altered. Knowing how to change a GUI's default appearance is an important set of knowledge to have on the job.

▶ Activity 1   GUI Customization and Multiple Windows

In this activity you will:

- Customize various properties of the GUI
- Use multiple windows within the GUI

Situation:

Ms. Jones is pleased with what you have taught her so far about how to use her "computer's windows". She was so enthusiastic about what she has learned that she subscribed to a computer magazine.

One of the articles she read related to "customizing the system's graphical user interface". Ms. Jones asks you to show her how to "customize" her system.

▼

Step 1—Review your software's on-line Help materials that relate to how to change the appearance of the GUI.

HELPFUL HINT

Read your on-line Help materials on how to change the features in Step 2. Record the default settings on your computer for each of the features listed in Step 2.

Step 2—Using the text editor or word processor, prepare step-by-step instructions for Ms. Jones on how to do the following:

a. Change the **default desktop color pattern** of the GUI

b. Change the **default background color** of the GUI

c. Change the icons which are displayed upon starting the GUI

d. Change the default **screen saver**

HELPFUL HINT

You will manage multiple windows and dialog boxes when changing the look of your GUI. Close one set of windows before attempting to change another feature.

Step 3—Test each of the instructions for accuracy.

Step 4—Print out a hard copy of the instructions.

Required Output: A printout of your instructions on how to perform each of the four tasks.

▶ **Activity 2   Controlling Output with the GUI**

In this activity you will:

- Change the default printer from within the GUI

Situation:

In her excitement, Ms. Jones also bought a color laser printer. She installed the printer interface card and a **printer driver** herself, but has asked you to show her how to change the GUI's **default printer** to the new color printer.

Step 1—Read the on-line Help materials that cover how to change the default printer from within the GUI.

Step 2—Prepare step-by-step instructions for Ms. Jones on how to change the default printer.

▼

activity 2 is continued on page 108

HELPFUL HINT

Set up the GUI printer settings so that a user can routinely switch from the more expensive color laser printer to a less expensive output device such as a dot matrix, ink jet or regular laser printer.

Step 3—Test your instructions for accuracy by altering the default printer from within the GUI. Produce a hard copy printout of the document.

Required Output: A hard copy printout on how to change default printers from within the GUI.

Activity 3   Altering the GUI System Font

In this activity you will:

- Change the GUI system's **default font**

Situation:

Ms. Jones asks you to change the default font used in the displays produced by the GUI. You tell her that it is possible. She asks you to prepare instructions for her on how to do it.

Step 1—Review the on-line Help materials on how to change the GUI's system font.

Step 2—Write step-by-step instructions that Ms. Jones can use.

Step 3—Test the accuracy of your instructions and print out a hard copy of the document.

Required Output: A hard copy printout of the instructions on how to change the GUI's default system font.

Other Activities and Questions

1. How many different default color schemes are associated with your system's GUI?
2. How many default screen savers are associated with your system's GUI?
3. Can you print out a file to a disk?
4. How many different printers can you choose from using your system?
5. Is there such a thing as screen savers that can be purchased and installed from outside sources and installed in your GUI?
6. Why are screen savers necessary?

Key Words and Phrases

default background color—The color or design of a GUI which is first displayed and later overwritten by windows and icons.

default desktop color pattern—The basic colors associated with a graphical user interface screen that have been preset either by the user or publisher.

default printer—The printer that would produce output unless directed to another device.

default system font—A preset font used by the GUI.

printer driver—A software program that makes it possible for a printer to work with a hardware system.

screen saver—A feature of a graphical user interface that causes screen output to be active.

Appendices
Summary of Skills to be Mastered

Appendix A Word Processing Skills

▶ Beginning Word Processing Skills

Activity 1—Creating a Document

1. Create and save a document that is **fully justified**
2. Center a title on a document
3. **Highlight** text
4. Change default text **fonts**
5. Change a font **style**
6. Use a **spell checker**
7. Produce a hard copy printout of the document

Activity 2—Retrieving and Modifying a Document

8. Retrieve a document from disk
9. Change font sizes
10. Use the word processor's **insert** and **type over mode**

Activity 3—Modifying Words and Phrases

11. Modify a document using the word processor's **thesaurus** feature
12. Modify a document using the **search** and **replace** feature of the word processor

Activity 4—Using Help and File Operations

13. Use the Save As option to store an existing document in **ASCII text format** under a different file name
14. Use the word processor's **Help** feature
15. Delete a document from storage

▶ Intermediate Word Processing Skills

Activity 1—Saving Files in an Alternate Format

16. Save a document in a different word processing file format

Activity 2—Specialized Retrieval of Documents

17. Retrieve a document from a drive and directory other than the default drive and directory
18. Edit a document that was saved in a different word processing file format

Activity 3—Creating a Boilerplate Document

19. Use **tab stops**
20. Change default margins
21. Create a boilerplate document

Activity 4—Creating a Newsletter

22. Center column headings
23. Use your word processor's basic text formatting tools
24. Use your word processor's **page numbering system**
25. Use the **line spacing** feature of your word processor
26. Change the orientation to your word processor's output

Activity 5—Blocking or Highlighting Selected Text

27. Use the block command
28. Delete text that has been blocked

▶ Advanced Word Processing Skills

Activity 1—Creating Documents with Graphics

29. Create a word processing document that contains a graphic
30. Use your word processor's drawing tools to create a simple image
31. Create a document that contains a **header** and **footer**
32. Add **borders** and **shading** to a portion of a document

Activity 2—Preparing a Form Letter

33. Use the word processor to prepare a form letter

Activity 3—Creating a Document with a Table

34. Create a document that contains a **table**

Activity 4—Merging a Data File

35. Create a data file to be merged with a form letter

Activity 5—Inserting Clip Art

36. Select a graphic from your word processor's clip art library and insert it into a document

Word Processing Skills Record Sheet

Student Name: _____

Description of Skills	Date Completed	Instructor Verification	Comments
Beginning Word Processing Skills			
Activity 1—Creating a Document			
1. Create and save a fully justified document			
2. Center a title on a document			
3. Highlight text			
4. Change default text fonts			
5. Change a font style			
6. Use a spell checker			
7. Print out a document			
Activity 2—Retrieving and Modifying a Document			
8. Retrieve a document from disk			
9. Change font sizes			
10. Use the insert and type over mode			
Activity 3—Modifying Words and Phrases			
11. Use the word processor's thesaurus			
12. Use the search and replace feature			
Activity 4—Using Help and File Operations			
13. Use Save As to store a document in ASCII format			
14. Use the word processor's on-line Help			
15. Delete a document			
Intermediate Word Processing Skills			
Activity 1—Saving Files in an Alternate Format			
16. Save files in a different word processor's format			
Activity 2—Specialized Retrieval of Documents			
17. Retrieve a file from an alternate drive and directory			
18. Edit another word processor's document			
Activity 3—Creating a Boilerplate Document			
19. Use tab stops			
20. Change default margins			
21. Create a boilerplate document			
Activity 4—Creating a Newsletter			
22. Center column heads			
23. Use basic formatting tools			
24. Use page numbering			
25. Alter line spacing			
26. Change the document's orientation			

Appendices

Page 113

Word Processing Skills Record Sheet

Student Name: _____

Description of Skills	Date Completed	Instructor Verification	Comments
Intermediate Word Processing Skills (continued)			
Activity 5—Blocking or Highlighting Selected Text			
27. Block text			
28. Delete blocked text			
Advanced Word Processing Skills			
Activity 1—Creating Documents with Graphics			
29. Create a document with a graphic			
30. Use the word processor's drawing tool			
31. Create a document with a header and footer			
32. Add borders and shading			
Activity 2—Preparing a Form Letter			
33. Create a form letter			
Activity 3—Creating a Document with a Table			
34. Create and insert a table in a document			
Activity 4—Merging a Data File			
35. Create a data file for merging			
Activity 5—Inserting Clip Art			
36. Insert clip art in a document			

Appendix B Spreadsheet Skills

▶ Beginning Spreadsheet Skills

Activity 1—Creating a Basic Spreadsheet

1. Enter **labels** and **values** in a **spreadsheet cell**
2. Construct and enter **formulas** in spreadsheet cells
3. Format numerical data as currency with two decimal places
4. Save a spreadsheet to disk
5. **Block** an appropriate **range** in a spreadsheet for printing
6. Print out a spreadsheet
7. Change default column widths

Activity 2—Retrieving and Modifying a Spreadsheet

8. **Copy the contents of a spreadsheet cell** to another cell
9. Insert a row and a column into an existing spreadsheet
10. Retrieve a spreadsheet
11. Access and use the spreadsheet's on-line Help feature

▶ Intermediate Spreadsheet Skills

Activity 1—Designing a Spreadsheet Template

12. Create a spreadsheet template
13. Format and **align** spreadsheet data
14. Use pre-defined spreadsheet functions

Activity 2—Modifying Templates and Selected Output

15. Modify an existing spreadsheet template
16. Print out a selected portion of a spreadsheet

Activity 3—Creating a Simple Spreadsheet Graphic

17. Create a simple bar graph from spreadsheet data

Activity 4—Inserting and Deleting Rows and Columns

18. Insert and delete a row of spreadsheet data
19. Insert a column in an existing spreadsheet

▶ Advanced Spreadsheet Skills

Activity 1—Modifying Cell Labels

20. Change cell label fonts
21. Change cell label parameters

Activity 2—Linking Spreadsheets

22. **Link spreadsheets**

Activity 3—Creating a Customized Graph

23. Create a **customized spreadsheet graphic**
24. Cut and paste data

Activity 4—Modifying Spreadsheet Output

25. Sort data in a spreadsheet column
26. **Export or print spreadsheet data to a file**
27. **Hide a column**
28. Use the Save As feature of a spreadsheet

Spreadsheet Skills Record Sheet

Student Name: _____

Description of Skills	Date Completed	Instructor Verification	Comments
Beginning Spreadsheet Skills			
Activity 1—Creating a Basic Spreadsheet			
1. Enter labels and values			
2. Construct and enter formulas			
3. Format numerical data			
4. Save a spreadsheet			
5. Block a range			
6. Print out a spreadsheet			
7. Change default column widths			
Activity 2—Retrieving and Modifying a Spreadsheet			
8. Copy cells			
9. Insert rows and columns			
10. Retrieve a spreadsheet document			
11. Use the spreadsheet's on-line Help feature			
Intermediate Spreadsheet Skills			
Activity 1—Designing a Spreadsheet Template			
12. Create a blank spreadsheet template			
13. Format and align spreadsheet data			
14. Use built-in spreadsheet functions			
Activity 2—Modifying Templates and Selected Output			
15. Modify an existing spreadsheet template			
16. Print out a selected portion of a spreadsheet			
Activity 3—Creating a Simple Spreadsheet Graphic			
17. Create a bar graph using a spreadsheet			
Activity 4—Inserting and Deleting Rows and Columns			
18. Insert and delete a row of spreadsheet data			
19. Insert a column in a spreadsheet			
Advanced Spreadsheet Skills			
Activity 1—Modifying Cell Labels			
20. Change the fonts of cell labels			
21. Change cell label parameters			
Activity 2—Linking Spreadsheets			
22. Link one spreadsheet with another			

Appendices

Spreadsheet Skills Record Sheet

Student Name: _____

Description of Skills	Date Completed	Instructor Verification	Comments
Advanced Spreadsheet Skills *(continued)*			
Activity 3—Creating a Customized Graph			
23. Create a customized graph			
24. Cut and paste data on a spreadsheet			
Activity 4—Modifying Spreadsheet Output			
25. Sort data in a spreadsheet			
26. Export a spreadsheet file to disk			
27. Hide a column in a spreadsheet			
28. Use the Save As feature of a spreadsheet			

Appendix C Database Skills

Beginning Database Skills

Activity 1—Designing and Using a Database

1. Design a database
2. Define a field
3. Specify a field's data type
4. Save a newly created database
5. Enter data using a **default format** to a data disk
6. **Sort records** in a database
7. Generate a report using a default format

Activity 2—Adding and Deleting Files

8. **Insert a new record or row** in an existing database
9. **Delete a record** or row in an existing database

Activity 3—Changing the Database's Structure

10. **Modify the structure** and contents of an existing database
11. Insert a column or field in an existing database
12. Delete a column in an existing database
13. Use the database software's on-line Help feature

Intermediate Database Skills

Activity 1—Using Key Fields and Calculating Fields

14. Create a **multitable database** with a **key field**
15. Create a **calculated field for a database**

Activity 2—Viewing Data and Correcting Errors

16. View data in a database
17. Identify and correct errors in database records

Activity 3—Renaming a Field

18. Rename a database field

Activity 4—Editing and Backing Up a Database

19. Back up data using a database program
20. **Edit** a particular record in a database
21. **Add a record** to a database
22. **Delete a record** from a database

Activity 5—Querying a Database and Using Logical Operators

23. **Query** an existing database
24. Use **logical operators**

Advanced Database Skills

Activity 1—Customizing a Data Entry Form

25. Develop a **customized data entry form**

Activity 2—Customizing a Report

26. Develop a **customized report format**
27. Create a report to be used as a form letter

Activity 3—Developing an Interactive Application

28. Develop an interactive database application

Activity 4—Importing, Querying and Exporting

29. Import a spreadsheet file into a database
30. Query a multitable database
31. Prepare a database file for export

Database Skills Record Sheet

Student Name: _____

Description of Skills	Date Completed	Instructor Verification	Comments
Beginning Database Skills			
Activity 1—Designing and Using a Database			
1. Design a database			
2. Define a field			
3. Specify a field's data type			
4. Save a database design			
5. Enter data using the default format			
6. Sort records in a database			
7. Generate a default report			
Activity 2—Adding and Deleting Records			
8. Insert a new record in a database			
9. Delete a record from a database			
Activity 3—Changing the Database's Structure			
10. Modify the structure of a database			
11. Insert a column in an existing database			
12. Delete a column in an existing database			
13. Use the database's on-line Help feature			
Intermediate Database Skills			
Activity 1—Using Key Fields and Calculating Fields			
14. Create a multitable database with a key field			
15. Create a calculated field for a database			
Activity 2—Viewing Data and Correcting Errors			
16. View data in a database			
17. Identify and correct errors in database records			
Activity 3—Renaming a Field			
18. Rename a database field			
Activity 4—Editing and Backing Up a Database			
19. Back up data using a database program			
20. Edit a particular record in a database			
21. Add a record to a database			
22. Delete a record from a database			
Activity 5—Querying a Database and Using Logical Operators			
23. Query an existing database			
24. Use logical operators			

Appendices

Page 121

Database Skills Record Sheet

Student Name: _____

Description of Skills	Date Completed	Instructor Verification	Comments
Advanced Database Skills			
Activity 1—Customizing a Data Entry Form			
25. Develop a customized data entry form			
Activity 2—Customizing a Report			
26. Develop a customized report			
27. Create a report to be used as a form letter			
Activity 3—Developing an Interactive Application			
28. Develop an interactive database application			
Activity 4—Importing, Querying and Exporting			
29. Import a spreadsheet file into a database			
30. Query a multitable database			
31. Prepare a database file for export			

Appendix D Presentation Graphics Skills

Beginning Presentation Graphics Skills

Activity 1—Creating a Presentation Graphics Slide

1. Access the presentation graphics on-line Help feature
2. Place text on a slide
3. Preview a slide
4. Spell check a slide
5. Alter the size, style and font of text that appears on a slide
6. Change the color of text on a slide
7. Relocate text on a slide

Activity 2—Using Clip Art

8. Insert a clip art image on a slide
9. Resize a clip art image
10. Change the location of a clip art image on a slide

Activity 3—Modifying a Slide and Printing a Copy

11. Copy and paste text on a slide
12. Copy and paste a graphics image to a slide
13. Alter an existing slide's orientation
14. Produce a hard copy output of a presentation graphics slide

Intermediate Presentation Graphics Skills

Activity 1—Creating a Master Slide

15. Create a **master slide**
16. Use the presentation graphics drawing tool to create an image

Activity 2—Creating a Properly Aligned Series of Slides

17. Design a series of slides communicating a concept
18. Properly align text on slides

Activity 3—Modifying Text Objects and Creating a Slide Show

19. Recolor a selected portion of an existing text object
20. Combine a series of screens into a slide show
21. Create a slide show that provides or manual slide changing

Activity 4—Editing a Slide Show

22. Edit an existing slide show

Advanced Presentation Graphics Skills

Activity 1—Creating Slide Show Special Effects

23. Add text to a **graphics object** on an existing slide
24. Create a bullet list slide upon which the individual items appear at different times
25. Vary the time that individual slides are available for viewing in a multi-screen slide show
26. Use various **transitions** in a presentation graphics slide show
27. Create a slide show with a logical branch

Activity 2—Altering Slide Images and Importing Graphics

28. Alter the background of an existing slide
29. Alter the color scheme of an existing slide
30. Alter the **shading** of an existing slide
31. **Import a graphics image** and insert it on an existing slide

Activity 3—Creating a Runtime Version Slide Show

32. Insert and incorporate a new slide into a previously saved slide show
33. Create a slide show with interactive logical branches
34. Produce a **runtime version** of a slide show

Presentation Graphics Skills Record Sheet

Student Name: _____

Description of Skills	Date Completed	Instructor Verification	Comments
Beginning Presentation Graphics Skills			
Activity 1—Creating a Presentation Graphics Slide			
1. Access the presentation graphic's on-line Help feature			
2. Place text on a slide			
3. Preview a slide			
4. Spell check a slide			
5. Alter the size, style and font of text on a slide			
6. Change the color of text on a slide			
7. Relocate text on a slide			
Activity 2—Using Clip Art			
8. Insert a clip art image on a slide			
9. Resize a clip art image			
10. Change the location of a clip art image on a slide			
Activity 3—Modifying a Slide and Printing a Copy			
11. Copy and paste text on a slide			
12. Copy and paste a graphics image on a slide			
13. Alter an existing slide's orientation			
14. Produce hard copy output of a presentation graphics slide			
Intermediate Presentation Graphics Skills			
Activity 1—Creating a Master Slide			
15. Create a master slide			
16. Use the presentation graphic's drawing tool to create an image			
Activity 2—Creating a Properly Aligned Series of Slides			
17. Design a series of slides communicating a concept			
18. Properly align text on slides			
Activity 3—Modifying Text Objects and Creating a Slide Show			
19. Recolor a selected portion of an existing text object			
20. Combine a series of screens into a slide show			
21. Create a slide show that provides for manual slide changing			
Activity 4—Editing a Slide Show			
22. Edit an existing slide show			

Appendices

Presentation Graphics Skills Record Sheet

Student Name: _____

Description of Skills	Date Completed	Instructor Verification	Comments
Advanced Presentation Graphics Skills			
Activity 1—Creating Slide Show Special Effects			
23. Add text to an existing graphics object			
24. Create a bullet list that builds			
25. Vary the viewing time between screens in a slide show			
26. Utilize a variety of slide transitions			
27. Create a slide show with a logical branch			
Activity 2—Altering Slide Images and Importing Graphics			
28. Alter the background color of an existing slide			
29. Alter the color scheme of an existing slide			
30. Alter the shading of an existing slide			
31. Import and insert a graphics image into a slide			
Activity 3—Creating a Runtime Version Slide Show			
32. Insert a slide into an existing slide show			
33. Create a slide show with logical branches			
34. Produce a runtime version of a slide show			

Appendix E Operating System Skills

Beginning Operating System Skills

Activity 1—Understanding Disk, Directory and Screen Operations

1. **Format a system disk**
2. Format a **data disk**
3. Produce a hard copy printout of a directory's contents
4. Change from the default directory
5. Clear the display screen

Activity 2—Using Essential File Operations

6. Display the directory's contents
7. Assess a file's characteristics
8. Switch the default drive
9. Change the name of a file

Activity 3—Accessing On-line Help

10. Access and use the operating system's on-line Help feature

Intermediate Operating System Skills

Activity 1—Copying a File

11. Copy a file from the hard drive to a disk

Activity 2—Creating Specialized Storage Space

12. Create a directory on the hard drive and disk
13. Copy a file from one storage space to another

Activity 3—Working with a Group of Files

14. Copy a **group of files** from one directory to another
15. Copy a group of files from one drive to another

Activity 4—Deleting Files and Specialized Storage

16. Delete selected files from a directory
17. Remove a directory

Activity 5—Adjusting the Time and Date

18. Change the operating system's time and date

Advanced Operating System Skills

Activity 1—Backing Up and Restoring

19. **Back up** selected files
20. **Restore** selected files

Activity 2—Modifying a File's Contents

21. Alter the contents of an ASCII file
22. Modify a key operating system file

Activity 3—Sorting Files

23. Sort files in a directory

Activity 4—Altering the System's Configuration

24. Alter the microcomputer system's configuration

Activity 5—Changing File Attributes

25. Alter individual **file's attributes**

Operating System Skills Record Sheet

Student Name: _____

Description of Skills	Date Completed	Instructor Verification	Comments
Beginning Operating System Skills			
Activity 1—Understanding Disk, Directory and Screen Operations			
1. Format a system disk			
2. Format a data disk			
3. Produce hard copy output of a directory's contents			
4. Change from the default directory			
5. Clear the display screen			
Activity 2—Using Essential File Operations			
6. Display a directory's contents			
7. Assess a file's characteristics			
8. Switch the default drive			
9. Change the name of a file			
Activity 3—Accessing On-line Help			
10. Access and use the operating system's on-line Help			
Intermediate Operating System Skills			
Activity 1—Copying a File			
11. Copy a file from the hard drive to a disk			
Activity 2—Creating Specialized Storage Space			
12. Create a directory on the hard drive and disk			
13. Copy a file from one storage space to another			
Activity 3—Working with a Group of Files			
14. Copy a group of files from one directory to another			
15. Copy a group of files from one drive to another			
Activity 4—Deleting Files and Specialized Storage			
16. Delete selected files from a directory			
17. Remove a directory			
Activity 5—Adjusting the Time and Date			
18. Change the operating system's time and date			
Advanced Operating System Skills			
Activity 1—Backing Up and Restoring			
19. Back up selected files			
20. Restore selected files			

Appendices

Operating System Skills Record Sheet

Student Name: _____

Description of Skills	Date Completed	Instructor Verification	Comments
Advanced Operating System Skills *(continued)*			
Activity 2—Modifying a File's Contents			
21. Alter the contents of an ASCII file			
22. Modify a key operating system file			
Activity 3—Sorting Files			
23. Sort the files in a directory			
Activity 4—Altering the System's Configuration			
24. Alter the microcomputer system's configuration			
Activity 5—Changing File Attributes			
25. Alter an individual file's attributes			

Appendix F   Graphical User Interface Skills

▶ Beginning Graphical User Interface Skills

Activity 1—Activating and Using the GUI

1. Initiate the computer's graphical user interface
2. Utilize a mouse for menu selections

Activity 2—Opening and Closing GUI Applications

3. Open and close a GUI application program

Activity 3—Becoming Familiar with the GUI

4. Identify the various parts of the GUI screen

Activity 4—Using the GUI's On-line Help

5. Use the on-line Help feature

▶ Intermediate Graphical User Interface Skills

Activity 1—Manipulating Windows

6. Open and close windows
7. **Resize** a window
8. **Reposition** a window
9. **Switch** from one GUI window to another

Activity 2—Formatting a Disk

10. Format a disk using the GUI

Activity 3—Using the Clipboard

11. Use the GUI's **clipboard**

Activity 4—Using a GUI Accessory

12. Use the calendar program that is associated with the GUI

▶ Advanced Graphical User Interface Skills

Activity 1—GUI Customization and Multiple Windows

13. Customize various properties of the GUI
14. Use multiple GUI windows

Activity 2—Controlling Output with the GUI

15. Change the default printer from within the GUI

Activity 3—Altering the GUI System Font

16. Change the GUI's default system font

Graphical User Interface Skills Record Sheet

Student Name: _____

Description of Skills	Date Completed	Instructor Verification	Comments
Beginning Graphical User Interface Skills			
Activity 1—Activating and Using the GUI			
1. Initiate the computer's graphical user interface			
2. Utilize a mouse for menu selections			
Activity 2—Opening and Closing GUI Applications			
3. Open and close a GUI application program			
Activity 3—Becoming Familiar with the GUI			
4. Identify the various part of the GUI			
Activity 4—Using the GUI's on-line Help			
5. Use the GUI's on-line Help feature			
Intermediate Graphical User Interface Skills			
Activity 1—Manipulating Windows			
6. Open and close windows			
7. Resize a window			
8. Reposition a window			
9. Switch from one GUI window to another			
Activity 2—Formatting a Disk			
10. Format a disk using the GUI			
Activity 3—Using the Clipboard			
11. Use the GUI's clipboard			
Activity 4—Using a GUI Accessory			
12. Use the calendar associated with the GUI			
Advanced Graphical User Interface Skills			
Activity 1—GUI Customization and Multiple Windows			
13. Customize various GUI properties			
14. Use multiple GUI windows			
Activity 2—Controlling Output with the GUI			
15. Change the default printer within the GUI			
Activity 3—Altering the GUI System Font			
16. Change the GUI's default system font			

Appendices